Peter W. Plummer

The Carpenters' and Builders' Guide

Being a Hand-book for Workmen, also a Manual of Reference etc. Seventh Edition

Peter W. Plummer

The Carpenters' and Builders' Guide

Being a Hand-book for Workmen, also a Manual of Reference etc. Seventh Edition

ISBN/EAN: 9783337059422

Printed in Europe, USA, Canada, Australia, Japan

Cover: Foto ©ninafisch / pixelio.de

More available books at **www.hansebooks.com**

THE

CARPENTERS' AND BUILDERS'

GUIDE,

BEING A

HAND-BOOK FOR WORKMEN,

ALSO A

MANUAL OF REFERENCE FOR CONTRACTORS, BUILDERS, ETC.

BY PETER W. PLUMMER.

SEVENTH EDITION.

NEW YORK:
W. T. COMSTOCK,
23 WARREN STREET.
1891.

Entered according to Act of Congress, in the year 1869, by
PETER W. PLUMMER,
in the Clerk's office of the District Court of the United States for the District of Maine.

PREFACE.

IN offering this little work to the public, the author feels that no apology is necessary. That such a work has long been needed, twenty years' actual experience as a carpenter is, to him, sufficient proof. Acting as he has during that time in a capacity where the wants of the carpenter are most apparent, he has endeavored to design and execute a work to meet those requirements. Great care has been taken in selecting materials for this important subject; and no article has been inserted, nor suggestion made, that is not of the highest practical value to workmen.

The first part treats upon building materials, and approximate method of making estimates, and may be depended upon when properly applied in estimating for materials of an average quality It is also recommended, that close attention be given to this subject, as no department of the business is of more importance, or requires more skill than this.

The glossary of architectural terms is full, and has been compiled from the best authorities.

The definitions and problems are of the highest importance to draftsmen, since without a thorough knowledge of them, but little progress can be made in the art. They have not been illustrated by diagrams, as it is believed the learner will obtain more practical information by following the directions given, and drawing the lines himself, than by any other method.

The principle illustrated on Plate I, fig. 3, also Plate III, figs. 3 and 4, is general in its application, and may be applied in the same way to all polygonal figures; also, with a slight variation, to many other problems in drafting.

The drawing for roof-cuts, Plate I, fig. 4, is a modification of the above principle, and a thorough knowledge of the same enables the carpenter to set bevels from his plan, for all the cuts necessary to be made in any framing operation, however complicated it may be.

The system of hand-railing, as given in this work, is the invention of the author, and is drawn from known properties of tangents and of the ellipse, and being thus mathematical in its character, it becomes general in its application, and applies with equal exactness to every case, however complicated. The process of drafting the same is also so natural, that all lines drawn by this system form angles that make the least possible variation from exactness.

It is also believed, that a thorough knowledge of this system will give the learner a broader view of the subject than any other and that carpenters, generally, will find it a valuable and interesting study.

With due regard for the interests of workmen, and a sincere desire for their advancement, and for a better understanding of the principles of their profession, to carpenters and builders generally, this little book is respectfully dedicated by

THE AUTHOR.

THE CARPENTERS' AND BUILDERS' GUIDE.

BUILDER'S ESTIMATES.

1. No definite rule can be given for making estimates; every case being peculiar to itself, and depending upon the size, style, and finish of the building to be erected, or job to be done. But to aid in this work, certain methods have been adopted, upon which it is safe to depend; and it is of the highest importance for carpenters and builders to be familiar with the process, as no contract for building can be safely made, without first making a correct estimate of all the materials necessary for the completion of the same.

And it is to assist in this work that we now propose to give the most approved methods of figuring, or making calculations of this kind; but it must always be borne in mind, that, notwithstanding all the assistance that can be derived from established rules, good judgment must be exercised in the selection of materials, in quality as well as quantity, as allowances for waste form an important item in all contracts, and depend, to a great extent, upon the good or bad quality of the materials to be used.

The following calculations are based upon the average quality of materials, and are, approximately, correct; yet, after all, much depends upon the experience and skill of the contractor, who, to be successful, must possess observation, energy, and good judgment; and, with these requisites, the cost of almost any job can be accurately determined.

BUILDING MATERIALS.

2. Building materials may be classified under several heads; but for our present purpose, only those used in masonry, carpentry, and the roofing, plumbing, etc., will be considered.

Of the first, the leading articles are stone, brick, lime, sand, gravel, hair, etc.; and of the second, timber, boards, furring, clapboards, shingles, laths, finishing boards, doors, sashes, blinds, hardware, etc., constitute the principal part. Now, as the cost of most of these articles, and others of equal importance, are included in nearly every estimate, and form an important item in every contract for building houses, stores, etc., it is important to know the method of estimating for the same.

We will, therefore, commence this subject by giving the customary method of measuring

STONE-WALLS, FOUNDATIONS, ETC.

3. The contents of stone-walls is estimated in perches, $24\frac{3}{4}$ cubic feet making one perch; therefore, to divide the cubic contents of a wall in feet by $24\frac{3}{4}$, gives the number of perches contained.

In walls eighteen inches in thickness, which are common, each $16\frac{1}{2}$ square feet of surface on the face of the wall makes one perch.

BRICKS.

4. Bricks may be estimated at 24 to a cubic foot, and five courses to one foot in height. But, as bricks are not often of

full size, the following allowances are made for each square foot of surface on the face of a wall, namely:

8 inch wall, 16 to a square foot.
12 " " 24 " " "
16 " " 32 " " "
20 " " 40 " " "

CHIMNEYS.

6. Bricks, for chimneys, may be estimated for each foot in height, as follows:

Size of Chimney.	Size of Flue.	Number of bricks to each foot in height.
16 × 16	8 × 8	30
20 × 20	12 × 12	40
16 × 24	8 × 16	40
20 × 24	12 × 16	45

CISTERNS.

7. A brick cistern may have a concrete bottom without a brick foundation, and, if built in a corner or against a wall, the adjoining part requires a wall only four inches thick, but the unsupported sides should be eight inches thick; hence, a cistern ten feet long, eight feet wide, and six feet deep, standing in a corner, requires 2736 bricks in its construction, while, if standing unsupported, 3712 will be necessary.

CESS-POOL.

8. A cess-pool, to receive and retain the sediment to be conveyed away in drains, should be small, that the least possible quantity of water may renew, and prevent its contents becoming stagnant.

CEMENT.

9. A cask of good cement will make into mortar sufficient for from 600 to 700 bricks, but is commonly mixed with sand, in the proportion of one cask of cement to two of sand.

In concrete floors, a cask of cement may be estimated for every nine square yards of surface, and should be mixed with gravel in the proportion of one cask to four of gravel.

LIME.

10. A cask of lime will make into mortar sufficient, commonly, for from 1000 to 1100 bricks; 1000 may be considered a safe estimate.

SAND.

11. Sand is estimated by the load; a load containing from 19 to 21 bushels. This is sufficient for about two casks of lime, therefore we may estimate one cask of lime to ten bushels of sand.

PLASTERERS' MORTAR.

12. Plastering mortar may be reckoned at about the same proportion of sand to a cask of lime, as brick mortar; but to make it tough and adhesive, one-half bushel of long hair, or three-fourths bushel of short hair should be mixed with each cask of lime.

A cask of lime is usually required for every 35 or 40 square yards of plastering surface.

Mastic is a kind of cement composed of sand, litharge, and linseed oil, and is used for plastering walls; the cost of which, for plain work, is about $1.25 per square yard.

FRAMES.

13. If sawed to dimensions, framing timber is liable to but little waste, though from ten to fifteen per cent is common.

In a large class of single, and in many double houses, and even in small blocks of dwellings, framing timber of the following dimensions is sufficient, and is much used, namely:

Sills,	7×8
Floor timber,	2×8
Posts,	4×6
Tie beams,	4×7
Studs,	2×4
Plates,	3×6
Rafters,	4×5
Studding for partitions,	2×3
Furring,	1×3

NOTE 1. Floor timbers, in small buildings, should be placed from 18 to 20 inches from center to center.

NOTE 2. Studs, in partitions and walls, are usually placed 16 inches from center to center; but in large buildings, 12 inches is the common distance.

NOTE 3. Furring, for ceilings, should be set off the same as studding.

NOTE 4. Rafters, in small buildings, are usually placed about 30 inches apart; depending much upon size and style of building.

NOTE 5. Upon these figures our present calculations are based, as there is a large class of buildings to which they will apply. But larger buildings require timber of larger dimensions; indeed, the variety is so great, that no general rule can be given.

BOARDS.

14. Boarding boards, under and upper floors, require from 15 to 25 per cent allowance for waste, and also for deficiencies in their measurement; while, for finishing lumber, from 33 to 40 per cent should be allowed.

CLAPBOARDS.

15. Clapboards are usually four feet long, and, when laid, from $3\tfrac{3}{4}$ to $5\tfrac{1}{2}$ inches of their width is exposed to the weather, depending upon the width of the clapboards.

We will confine our estimate to 4, $4\tfrac{1}{2}$, and 5 inches exposed to the weather; hence,

 4 inches requires 75 to 100 square feet.
 $4\tfrac{1}{2}$ " " $66\tfrac{2}{3}$ " " "
 5 " " 60 " " "

SHINGLES.

16. Shingles are usually 16 inches long, and a bundle of shingles is 20 inches wide, and contains 24 courses in thickness at each end; hence, a bundle of shingles will lay one course 80 feet long.

Shingles, upon the roof of buildings, should never be exposed to the weather more than five inches, nor upon walls more than six inches; therefore, when shingles are exposed,

 4 inches, 1000 will cover 107 square feet.
 $4\tfrac{1}{2}$ " " " " 120 " "
 5 " " " " 132 " "
 6 " " " " 160 " "

LATHS.

17. One bundle of 100 laths is estimated to cover $5\tfrac{1}{2}$ square yards of surface; also, 1000 laths to 55 square yards. Consequently, one cask of lime will make into mortar sufficient to spread about the same surface as 700 laths.

NAILS.

18. For 100 clapboards, allow about 3½ lbs. of 5d. nails.

For 1000 shingles, from 5½ to 6 lbs. 4d. nails; or, from 3½ to 4 lbs. 3d. nails.

For 1000 laths, allow 7 lbs. 3d. fine nails.

Name.	Length.	Number to a pound.
3d. fine,	1 inch,	557
3d.	1¼ "	385
4d.	1⅜ "	254
5d.	1½ to 1¾ inch.	232 to 180
6d. finish,	2 "	215
6d.	2 "	154
7d.	2½ "	141
8d. finish,	2½ "	150
8d.	2½ "	90
9d.	2¾ "	76
10d. finish,	3 "	84
10d.	3 "	62
12d.	3¼ "	50
20d.	3⅝ "	36
30d.	4 "	24
40d.	4½ "	18
50d.	5⅛ "	13
60d.	6 "	9
70d.	7 "	6

NOTE. The nails of different manufacturers vary somewhat in size but not enough to lessen the uses of the table.

SLATES.

19. The following table gives the number of slates of different sizes required to cover a square of 100 feet, for a lap

of two inches, and also for three inches; the laps of 2¼, 3½, and 4 inches not being given, as they are not often required.

Size.	Lap 2 inches.	Lap 3 inches.
12 × 6	480.00	533.33
7	411.43	457.14
8	360.00	400.00
14 × 7	342.86	374.03
8	300.00	327.27
9	266.67	290.91
10	240.00	261.82
16 × 8	257.14	276.92
9	228.57	246.15
10	205.71	221.53
11	187.01	201.40
18 × 9	200.00	213.33
10	180.00	192.00
11	163.64	174.55
18 × 12	150.00	160.00
20 × 10	160.00	169.41
11	145.45	154.01
12	133.33	141.17
13	123.08	130.32
14	114.29	121.01
22 × 11	130.91	137.80
12	120.00	126.32
13	110.77	116.60
14	102.86	108.27
15	96.00	101.05
16	90.00	94.74
24 × 12	109.09	114.29
13	100.70	105.57

14	93.51	97.96
15	87.27	91.43
16	81.82	85.71
17	77.01	80.67
18	72.73	76.19

Hence, for a lap of two inches it requires 90 slates, 22×16, to cover one square.

TIN FOR ROOFING.

20. Our estimate for tin will be for two sizes only, namely, 10×14 and 14×20: of the first, it requires 123 sheets to cover one square; and of the second, 59 sheets will cover 101 square feet.

WOOD SCREWS.

21. Wood screws vary in size from 1-16 to $\frac{3}{8}$ of an inch in diameter, and in length from $\frac{1}{4}$ to $3\frac{1}{2}$ inches, and include 23 numbers, thus:

Length.	Number.	Length.	Number.
$\frac{1}{4}$	0 to 2	$1\frac{1}{2}$	9 to 18
$\frac{3}{8}$	2 to 7	$1\frac{3}{4}$	10 to 20
$\frac{1}{2}$	2 to 9	2	11 to 20
$\frac{5}{8}$	3 to 10	$2\frac{1}{4}$	11 to 20
$\frac{3}{4}$	4 to 12	$2\frac{1}{2}$	14 to 20
$\frac{7}{8}$	6 to 12	$2\frac{3}{4}$	16 to 20
1	7 to 14	3	15 to 20
$1\frac{1}{4}$	8 to 16	$3\frac{1}{2}$	18 to 22

NOTE. No 11 is about 3-16, and No. 14 about $\frac{1}{4}$ of an inch in diameter, and are sizes in very common use.

SAND PAPER.

22. Sand paper is numbered as follows:

No. 0, finest.
" $\frac{1}{2}$, "
" 1, "
" 1$\frac{1}{2}$, medium.

No. 2, medium.
" 2$\frac{1}{2}$, "
" 3, coarsest.

BUTTS AND LOCKS.

23. To determine whether doors require right or left-handed butts and locks, always imagine the door swinging from you; and if to the right, it requires right-handed butts and locks, but if to the left, the reverse.

LEAD PIPE.

24. The weight of lead pipe may be estimated for each foot in length, as follows:

$\frac{3}{4}$ inch bore, 1 lb. 12 oz.
1 " " 2 " 5 "
1$\frac{1}{4}$ " " 2 " 12 "

1$\frac{1}{2}$ inch bore, 4 lbs. 5 oz.
2 " " 5 " 10 "

IRON SINKS.

25. Iron sinks vary in size from 1$\frac{1}{2}$ × 2$\frac{1}{2}$ to 2 × 4 feet.

SHEET LEAD.

Thickness.	Weight per square foot	Thickness.	Weight per square foot.
1-16 inch,	4 lbs.	1-8 inch,	7$\frac{1}{2}$ lbs.
1-10 "	6 "		

TABLE

OF ALLOWANCES MADE BY PLASTERERS FOR DOORS AND WINDOWS.

26. For doors in partitions when both sides are plastered, one side is to be deducted; and for windows and outside doors, one-half the surface of each is to be allowed.

The table exhibits the surface of the following sizes of doors and windows, which will be convenient for reference when measuring plastered surfaces.

DOORS.

SIZES.	SURFACE IN FEET.	SIZES.	SURFACE IN FEET.
Ft. In. Ft. In.		Ft. In. Ft. In.	
2 6 × 6 6,	17	2 10 × 6 10,	21
2 8 × 6 8,	18	3 0 × 7 0,	22

WINDOWS.

With twelve lights in each window, and glass of the following sizes, namely:

Size of glass.	Surface in feet.	Size of glass.	Surface in feet
9 × 12 inches,	12	11 × 16 inches,	18
9 × 13 "	14	12 × 20 "	25
10 × 14 "	17	12 × 24 "	29

EXAMPLE.

A room 13 × 15 ft., and nine feet in height, with five doors 2 ft. 10 in. × 6 ft. 10 in.; and three windows of twelve lights each, of 12 × 20 inch glass, contains 67$\frac{1}{2}$ yards of plastered surface.

WINDOW FRAMES.

Window frames for twelve lighted windows should be five inches longer than four lights, and 3$\frac{5}{8}$ inches wider than three; and for four lighted windows, the length should be 5¼ inches longer than two lights, and 3⅞ wider: these allowances are made for stiles, rails, mullions, etc.

Outside blinds should be ½ inch longer than the sashes, and of the same width.

COAL BINS.

27. A coal-bin to hold one ton of anthracite coal must have a capacity of at least 36 cubic feet; for two tons, twice that number of feet, etc.; hence a coal-bin five feet square at the bottom, and five feet deep, will hold about four tons of coal.

CISTERNS.

28. The capacity of a cistern is estimated in gallons; 7½ gallons to a cubic foot, which though not strictly correct, is near enough for practical purposes; hence, to get the contents of a cistern in gallons, multiply the product of the length, breadth, and depth of the inside, in feet, by 7½.

FOR EXAMPLE.

A cistern ten feet long, eight feet wide, and six feet deep on the inside, contains 3600 gallons.

WOODEN CISTERNS.

Are commonly shaped like the frustrum of a cone, and as they are in common use, we will give the rule for measurement:

First, find the surface of the two bases, and multiply them together, and extract the square root of the product; then add to this root the two bases, and multiply this sum by ⅓ the height of the cistern; this gives the cubic feet in the cistern, which being multiplied by 7½, gives the number of gallons contained.

NOTE 1. A cubic foot contains 1728, and a gallon 231 cubic inches; hence 1728 divided by 231 gives 7½ nearly, or number of gallons in a cubic foot.

NOTE 2. A bushel contains 2150.42 cubic inches.

NOTE 3. A barrel is about 2½ feet high, and about 20 inches in diameter at the bilge.

ESTIMATES.

29. It often becomes necessary for the carpenter or builder to make what is called a "rough estimate" of the amount of materials required to construct a building, oftener, perhaps, for cheap buildings than others; we will, therefore, add two examples, which will assist in making similar estimates.

BUILDING.

FIRST EXAMPLE.

House 22 × 34 feet, posts fifteen feet in height. Addition to same 16 × 20 feet; posts nine feet in height. It may be well to add, that twenty-two by thirty feet is the size usually adopted for the one and one-half storied house, so common in all parts of the country; but the former dimensions have been selected, because an actual occurrence. This building was finished in the usual style, and required, among the leading items, as follows:

Timber, allowing usual waste,	3800 feet.
Furring and studding, 1 × 3, and 2 × 3,	1100 "
Outside boarding, including 15 per cent waste,	3100 "
Under floors, " " " "	1750 "
Upper floors, " " " "	1750 "

Clapboards, 4½ in. to weather, about . . 825 in No.
Shingles, 5 in. to weather, . . . 9¼ M.
Laths, ... 9000 in No.
Plastering, ... 562 yards.
Board nails, ... 225 lbs.
Floor nails, ... 60 "
Clapboard nails, ... 27 "
Shingle nails, ... 50 "
Lath nails, ... 60 "

ADDITION TO THE SAME.

16 × 20 feet, posts nine feet in height.

Timber, ... 1225 feet.
Furring and studding, ... 250 "
Boarding boards, ... 1200 "
Under floors, ... 400 "
Upper floors, ... 400 "
Clapboards, ... 275 in No.
Shingles, ... 4 M.
Laths, ... 1800 in No.
Plastering, ... 110 yards.
Nails, etc., omitted.

SECOND EXAMPLE.

House 22 × 34, posts 22 feet in height.
Addition 16 × 20 feet, posts 22 feet in height.

This is the common two-storied house, with finished attic, and is the most convenient and economical house in the country. Estimating as above, it requires as follows:

Timber, ... 5000 feet.
Furring and studding, ... 1500 "
Boarding boards, ... 3900 "

THE CARPENTERS' AND BUILDERS' GUIDE. 19

Under floors,	2500 feet.
Upper floors,	2500 "
Clapboards, in number,	1050
Shingles,	9¼ M.
Laths,	15000 in No.
Plastering,	850 yards.
Board nails, commonly,	300 lbs.
Clapboard nails,	35 "
Shingle nails,	50 "
Lath nails,	100 "

ADDITION TO SAME.

16 × 20 feet, posts 22 feet in height.

Timbers,	2100 feet.
Furring and studding,	425 "
Boarding boards,	2000 "
Under floors,	750 "
Upper floors,	750 "
Clapboards, in number,	600
Shingles,	4 M.
Laths,	3500 in No.
Plastering,	225 yards.

These examples are given to assist in making "rough estimates;" but in the construction of buildings it is necessary to make a careful estimate of each and every article entering into them, and allowances must be made, generally, in proportion to the quality of the materials to be used; the first quality, in nearly all cases, requiring much less allowance for waste than inferior qualities.

DURABILITY OF SHINGLES.

30. The following table exhibits the average durability of shingles in exposed situations:

    Rifted pine shingles from 20 to 35 years.
    Sawed, clear from sap, " 16 to 22 "
    "   " with " " 4 to 7 "
    Cedar shingles,   " 12 to 18 "
    Spruce, "   " 7 to 11 "

NOTE. By soaking shingles in lime-water, their durability is considerably increased.

31. COST OF DIFFERENT KINDS OF ROOFING.*
    Shingles, from 5 to 12 dollars per square.
    Slates,   " 14½ to 17 " " "
    Tin,   " 13 to 15 " " "
    Gravel composition, 7 " " "

PAINTING.

32. No general rule can be given for estimating the cost of painting, the surfaces to be painted, and the conditions otherwise being so different; but it may be proper to state, that from twenty to twenty-five cents per square yard is charged for two coats on the outside of new buildings, the extra expense on windows generally balancing the "outs." Furthermore, that from four to six gallons of oil is required for each 100 pounds of white lead, depending also upon circumstances.

---

* Cost in 1869.

PROBLEMS FOR DRAFTING.

BEFORE proceeding with this subject, it is necessary to procure appropriate instruments for drafting purposes. These consist of a T square, triangle, dividers, pencils, eraser, and some drawing paper. Furthermore, it is important for the learner to be exact in his drawings, and also to contract a habit of neatness, as this qualification is indispensable to the good workman.

To divide a Straight Line into two equal parts.

Draw a horizontal line of convenient length; denote the left by A, the right by B; then extend the dividers, say two-thirds its length, and with first A, and then B for a center, make intersections above and below the line; then a line passing through the intersections, and cutting the line A B, will divide it into two equal parts.

At a given point in a given Straight Line to erect a Perpendicular to the Line.

Take the same line A B, as before, and denote the given point on the line by C; then from the point C, to the right and left, set off equal distances; denote these points by D and E (both to be on the line, and inside of A B); then, with the points D and E for centers and a radius greater than the distance C D or C E, describe two arcs, intersecting above the line; then a line drawn through this intersection, and through the point C, will be perpendicular to the line at C.

NOTE. This construction serves to draw a right angle at a given point.

From a given point, without a Straight Line, to let fall a Perpendicular on the Line.

Denote the line by A B, and the given point above the line by C. Then, with C for a center and a radius sufficiently great, describe an arc, cutting the line A B at two points; denote these points by D and E, then first with D for a center, and then E, and a radius sufficiently great, make an intersection below the line, or on the side opposite the point C; denote the intersection by H, then a line drawn through C and H will be a perpendicular from C to the line A B.

To divide an Arc or Angle into two equal parts.

First, draw a line, which denote by A B; then draw a line from A and one from B, intersecting above the line,—these form a triangle. Denote this triangle by A, B, C; let us now bisect the angle A. If A B be the shortest side that makes the angle A, set off the distance A B on A C; but if A C be shortest, set off A C on A B,—in either case denote the point by D. Then, first with D for a center and then B (if A B be the shortest line), make an intersection outward from the angle A, to the right, with the same radius A B; then a line drawn from A through this intersection, divides the angle B A C into two equal parts. If the angle be subtended by an arc of a circle, instead of a straight line, the same principle will apply.

Through a given Point, to draw a line parallel to a given Straight Line.

Draw a horizontal line of convenient length, which denote by B C; then denote by A the given point, which place at a convenient distance above B C; then with A for a center and a radius greater than the shortest distance to the line B C, describe an indefinite arc, from a level with A, to the line B C; denote the intersection on the line B C by E; then, with E for a center and the same radius, describe another arc, cutting the line B C and the point A; denote the point of intersection on the line B C by F; then set off on the first arc the distance A F, and denote the point by D; then a line drawn from A through D, will be parallel to the line B C.

To find the Center of a given Circle.

Denote by A, B, and C any three points on the circumference. Connect A to B, and B to C, by straight lines; then find the center of each line, and make a point on each; then from each point let fall a perpendicular to the line toward the centre of the circle, and the point of intersection is the center of the circle.

NOTE. This principle applies when three points are given to find the centre of a circle, whose circumference will intersect all of them.

Three sides of a Triangle being given to describe the Triangle.

First, describe the longest side, which denote by A B; then extend the dividers to the length of the second side, and with A for a center, describe an arc; then extend the dividers to the length of the shortest side, and, with B for a center, describe another arc; then from the point where these arcs intersect draw lines, one to A and the other to B, and we have the triangle required.

NOTE. This principle applies in describing an equilateral triangle, the three sides being equal.

To inscribe a Square in a given Circle.

Draw two diameters to the circle, at right angles to each other, and join their extremities, and these lines are the four sides of the inscribed square.

In a given Circle to inscribe a regular Hexagon, and also an Equilateral Triangle.

First apply radius, or half the diameter six times to the circumference, which brings us round to the point of beginning; then connect these points by straight lines; each line is a side of the inscribed hexagon.

To inscribe the Equilateral Triangle.

Denote the six points on the circumference by A, B, C, D, E, and F, and connect A to C, C to E, and E to A; then A C E is the inscribed equilateral triangle.

To reduce a Square to an Octagon.

Describe the square and find its center, then draw the circumscribing circle. Find the centre of each side of the square; draw two diameters to the circle through these points; connect the extremities of these diameters, forming another square. Then the projecting corners being gauged on the square, show what must be taken off to form an octagon.

Another method to obtain the side of an octagon is to draw a right-angled triangle. Let its base be twelve inches in length, and its perpendicular five inches in height; then draw the hypothenuse, or connecting line. Now to find the side of an octagon, to be cut from a square, set off on the base of the triangle a side of the square, and draw a perpendicular from this point to the hypotenuse,—the length of this perpendicular is a side of the required octagon.

To find the Segment of a Circle, the Center being unknown.

Draw a cord of any convenient length, and denote it by D E. Denote the center of the cord by A, then measure the distance from A to the highest point of the arc, and denote it by B; then, with A for a center, and a radius A B, describe a quadrant between A B and A E. Let the quadrant be denoted by A B C. Now divide the arc of the quadrant into any number of equal parts, say six; then divide its radius A C into the same number of parts, and join the points of division; then divide A E and A D into the same number of parts, and draw lines to correspond in their angles to those in the quadrant, and their length being transferred to these lines, give points by which to describe the curve.

To describe a regular Pentagon, and also a regular Decagon.

First draw a circle of convenient size, and also a diameter to it. Denote the center of the diameter by C, and one extremity by A, and at the point A erect a perpendicular equal in length to one-half of C A. Denote the upper extremity of this perpendicular by B, and connect B and C; then, with B for a center and a radius B A, describe an arc intersecting the line C B. Denote the intersection on C B by D, then C D is a side of the decagon, and requires to be applied ten times to the circumference to arrive at the point of beginning; then by joining alternate corners we have the regular pentagon, or figure with five equal sides.

To describe an Ellipse, or Oval.

Draw a line of the same length as the major axis, or greatest diameter, which denote by C C, Plate IV, fig. 3; then find its center and draw one-half its minor, or less, axis perpendicular to it. Denote the center by A, and the upper extremity of the minor axis by B; then extend the dividers from A to C, and with B for a center, make intersections on the major axis to the right and left, at two points, which denote by F and F; then stick a pin at F, one at B, and one at F on the right; then make a loop in the end of a linen thread, and make one end fast at F, and bring it round the point B to F on the right, and wind it two or three times around the pin at this point (remember the thread must be kept drawn during the process); then remove the pin from B, and with a pencil (notched to prevent slipping) placed in the bight at *h*, and the thread kept drawn, proceed to describe the ellipse, or oval.

To divide the width of a Board into any number of equal parts.

For example, let the board be eight inches wide; place a rule across it at such an angle as will bring an end even with each side; then if the board is to be divided into six equal parts, by marking every space of two inches on the rule, we have the points of division.

NOTE. This rule applies in making divisions for flutings on pilasters, and other work.

CUTS IN CARPENTRY.

To describe the cuts for a board whose lower edge, when in position, shall be horizontal or level, and whose face shall be curved, and also incline from a perpendicular, as the back of a curved pew, the front of a gallery, etc.

Denote the curve to which the board is to be bent when in position by A B, and by rules already given find the radius to this curve. Denote the radius by 2, 3; then from 3 draw a line perpendicular to 2, 3, and from 2 draw a line on the same slope, or inclination, as the board when in position, to an intersection with the perpendicular line from 3. Denote this point by C, then set off from 2, toward C, the width of the board, and denote the point by D; then, with C for a center and a radius C 2, describe an arc. Again, with C for a center and a radius C D, describe another arc, and make it of the same length of the arc A B; then a line drawn from C through its extremity, and through the greater arc, gives the lengths of the two arcs, giving a pattern of the required form. Denote the last intersections by H and J, then D 2 H and J is the form required.

CUTS FOR BOXES OR HOPPERS WITH FLARING SIDES.
Plate III.

For the first example, let it be required to set bevels for the cuts for a hopper, whose top and bottom are two unequal squares, or rectangles.

First, draw the square A, B, C, D, Plate III, fig. 3; through C draw C E to the same angle with C D that side of the box when completed makes with a perpendicular line; then let fall a perpendicular to C E, from F through D. Also, draw H E parallel with C D; then, with C for a center and a radius C F, make an intersection at L, and connect A and L, and at A is the bevel for the mitre at the ends. Again, with D for a center and a radius D F, make an intersection at K, and connect B and K, and at B is the bevel for the down or cross cut. Sometimes it may be desirable to cut the ends square as possible, instead of cutting by a mitre; in this case make E a center, and with E F for a radius, make an intersection at I, and connect I and H, and at II is the bevel for the square cut of the ends.

As a second example, let it be required to set bevels for the cuts for a box whose top and bottom are two unequal equilateral triangles, with flaring sides.

In this example, first draw the equilateral triangle C E F, Plate III, fig. 4, by rules already given; then from the middle point of E F, let fall the perpendicular B D; then draw A B parallel and equal to C D, also A C parallel and equal to B D, thus forming the rectangle A, B, C, D. Now draw C H to the same inclination from C D, that a side of the box, when completed, is from a perpendicular line; then draw L D perpendicular to C H, and, with C for a center and a radius C L, make an intersection at I, and connect I and A, and at A is the bevel for the mitre at the ends. Again, with D for a centre and a radius D L, make an intersection at K, and connect K and B, and at B is the bevel for the down, or cross cut.

For the last example, let it be required to set bevels for the cuts for a box, whose sides incline from a perpendicular, and whose top and bottom are two unequal hexagons, or six-sided figures. In this case, draw a horizontal line C E, Plate I, fig. 3, and find its center D; then raise the perpendicular D B indefinitely, and, with E for a center and

C E for a radius, make an intersection at B; then connect E and B and draw A B parallel and equal to C D; also, A C parallel and equal to B D, again forming the rectangle A, B, C, D, as in the two last examples. Now draw C H to the same angle with C D, that a side of the box makes with a perpendicular line; then draw L D perpendicular to C H. and, with C for a center and a radius C L, make an intersection at K, and connect A and K, and at A is the bevel for the mitre of the ends. Again, with D for a center and a radius D L, make an intersection at I; also connect I and B, and at B is the bevel for the down or cross cut.

It is not necessary to give other examples for the cuts for boxes with unequal polygonal tops and bottoms, as all examples of this kind come under one rule; but it should be observed, that when the bevels are applied, the edges of the stock or boards must, in all cases, be square.

RAKING MOLDINGS.

Plate IV.

To draft raking moldings that will compare with level moldings; also, to make the cuts across a box by which they can be mitred.

Draw the square figure, A, B, C, D, Plate IV, fig. 2; then draw C E to the same rake or pitch as the raking molding when in its place. Also draw C b, the outline of the level molding, in the same position as if in its place; then draw 1 b perpendicular to C D. Also, draw 1, 6 square with 1 b. Now draw from 1, 6 as many perpendiculars intersecting the level molding at prominent points as may be convenient; then with 1 for a center, and the radii 1, 2; 1, 3; 1, 4, etc., make intersections on C E, where corresponding numbers are seen. Also, draw b d parallel with C E; then from the intersections on the molding, C b, draw lines parallel with C E, to an intersection with 6 d; then from the intersections at 1, 2, 3, etc., on the line C E draw lines perpendicular to it, and the intersections with the lines from C b give points by which to describe the raking molding, 6 m* being its outline.

Now to make the cuts across a box by which this molding can be mitred, first draw D F perpendicular to C E; then, with C for a center and a radius C F, make an intersection at H, and connect A and H,

---

* In all cases, the intersections are the points through which the outline of 6 m must be drawn, but through the carelessness of the engraver this is not the case in this drawing, the curve starting from 6, instead of the intersection with the first line below, and terminating at m, instead of the intersection with the first line above. The remainder is drawn correctly.

and at A is the bevel for the cross cut; also, the bevel for the down cut is seen at E, while the square mitre at B applies to the level molding C b. Also, observe that the line 1 m on the raking molding should be perpendicular against the side of the box, when in position for cutting.

In the above example 6 m is the outline of a raking molding, that will compare with the level molding C b, and if the level molding be cut by a square mitre, the cross and down cuts of the rake must be made by the bevels at A and E respectively. Now it sometimes happens, that it is necessary to mitre another level molding to the upper end of the rake; in this case, the down cut of the rake must be made by the bevel at E, and the cross cut by a square mitre. Consequently, it becomes necessary to draft another molding, that when level will compare with this section, or outline of the cut, on the rake; in this case the raking molding may be assumed to be level, and by applying the above method, a correct outline of the required level molding is obtained; and, furthermore, all moldings thus drafted will compare with a level molding, if cut to any angle, as well as a square mitre.

ROOF CUTS, INCLUDING JACK RAFTERS.

Let A B, A C, Plate I, fig. 4, represent the center of the wall-plates forming the angle of a building; then bisect this angle, giving A D for the seat of the hip rafter, or the line over which its center will stand. Also, draw C D produced square with A C for the seat of a jack rafter; then draw D E perpendicular to C D, to the same height that it will stand when in its place. Also, draw C E; this gives the length of a jack rafter and also the bevel for the down cut at E. Now, with C for a center and a radius C E, make an intersection at F, then connect A and F, and at F is the bevel for the cross cut of the upper end; also, C is the bevel for the end resting on the plate.

The above lengths can be obtained from the plan, and should be proportioned by it, a convenient size being 1½ inches to a foot for an ordinary job.

BACKING A HIP.

For a second example, let A B, A C, Plate II, fig. 3, represent the center of the wall-plates, then bisect the angle giving A D for the seat of the hip, and raise E D perpendicular to it, to the height of the hip rafter when in its place; also, connect E and A, and E A is the length of the hip rafter; also, the angle at A gives the bevel for the foot cut of the same. Now, backing a hip signifies to take off its corners so that when in its place the upper side will range with the rafters; for this purpose make A a equal A b, and connect a and b; then, with o for a center and a radius touching A E, make an intersection at d, then connect d to a and b; then a bevel set to the angle a b d properly applied, gives a direction for backing the hip, as the form of the top must be a d b, when in range with the rafters.

TO DESCRIBE HIP OR GROINS
FOR THE INTERSECTION OF ARCHED OR CURVED CEILINGS.

Draw A B, A c, Plate II, fig. 2, square with each other, and set off on A B the projection of the arch or cornice on the ceiling; also, make A c equal to its height or length on the wall below the ceiling. Then place one of the ribs, as c a, in the same position with respect to A B, A c, that it would have with the wall and ceiling, if in its place. Then draw lines parallel with A B, from a, b, c, d, etc., through prominent points on the rib as often as necessary; then if the hip or groin is to be placed in a square corner, measure the distance that any intersection of the rib may be from the vertical line A c; and for each foot set off 17 inches* on the intersecting line, or in that proportion. Thus if a a be 20 inches in length, then make a point at d, 28$\frac{1}{3}$ inches from a; make a similar calculation for each point, then connect the points thus obtained, and we have d N, or the mold for the hip.

If the intersections do not form a square corner, then the proportion of allowance can be obtained by drawing two lines to the same angle as the corner, as A B, A C, Plate I, fig. 4. Here C D represents a projection on the ceiling; then A D, or the seat of the hip rafter, is the proportion of allowance to be made for the groin. In this case, the proportion of allowance is as D C is to A D, and a similar calculation can be made for all other angles.

---

* This rule applies when setting off for braces, as the distance from corner to corner of any figure 12 inches square is 17 inches; consequently, if the figure be three feet square, the distance is three times 17 inches, or four feet, three inches.

STAIR RAILS.

First Example.

The only difficulty in the art of stair railing consists in describing a wreath-piece that will stand over a circle, on a given inclination from a level. Now an ellipse is the only curve that will satisfy this condition, and when properly described will stand exactly over any required circle or its arc; but, in order to do this, the following conditions are required to be known, namely: 1st, the diameter of the cylinder from center to center of the balusters; 2d, the pitch of the straight rail, or that connecting with the wreath; 3d, the height of the upper, above the lower end of the same; and 4th, the width and thickness of the rail. Now, in order to find the required parts, it is necessary to draw a plan of the stairs, also the cylinder or well-hole, over which the wreath-piece is required to stand; furthermore, it is convenient to work from the center of the rail, consequently any line representing the pitch of a rail is supposed to be in the position the center of the rail would occupy if the bottom of the same was resting on the steps; and when reference is made to the joints to each end of the wreath, the center of the joint, or where the line, passing through the center of the rail, intersects the joint, is the point that is referred to.

Before proceeding further, it may be necessary to explain some of the properties of tangents, as a correct understanding of the art depends much upon them.

Tangents are straight lines touching a curve at only one point, and the properties of a tangent depend upon the nature of the curve with which it is in contact. A tangent to a circle is the most simple, and the only condition required is, that it form a right angle with a diameter of the same. A tangent to an ellipse has more complicated properties; it has a point of contact with the circumference, but does not form a right angle with a diameter, as in the circle, except at the extremities of the major and minor axis. As an example, let us suppose a circle drawn, and inclosed within a square; then the sides of the square are tangents to the circle. Now an ellipse is the only curve that will stand over a circle, on an inclination from a level, hence the tangents to the ellipse must also stand over those to the circle. From this property of tangents we are enabled to determine what arc of an ellipse will stand over any required arc of a circle.

But first, in order to determine the parts mentioned above, et A C Plate I, fig. 1, represent a quarter of a circle, or cylinder over which

THE CARPENTERS' AND BUILDERS' GUIDE. 31

the center of a wreath-piece is required to stand; then the curve A C is the line passing through the center of the balusters, and A B, B C are tangents to it, because there is a point of contact with each at A and C ; and they also form a right-angle with the radii A O, C O, respectively. And as the curve that will stand over A C will be limited by tangents, they must also stand over A B, B C; consequently, as A B, B C are respectively equal to A O, C O, or a diameter of the cylinder, they may be spread out at C A B, fig. 2.

Also, the pitch of the straight rail can be obtained from a step and riser, or horizontal distance over which the rail passes, compared with its height; consequently, it can be taken from the elevation, as can also the height of the center of one end of the wreath-piece, above the center of the other; furthermore, the width and thickness of the rail is optional, and is governed by the style and finish of the stairs. Now, having determined these conditions, extend B A to C, C O to A, and M X to B, fig. 2, and from C draw C D to the same pitch as the straight rail. Also, let C W equal the height of the center of the joint over C, above the center of the joint at A, then draw W E parallel with C B and produce C D to H; also, draw D E, produced to n. Now C D and D E are the tangents to an ellipse that will stand over A C, and it is required to arrange them in a form to stand over A B and B C, fig. 1. For this purpose make J E equal C D, and let fall the perpendicular J t; then raise a perpendicular to n E, from A to P; also, let fall a perpendicular to C H through I; then, with I for a center and a radius I M, make an intersection at L, and draw a line from E through L to K; then, with E for a center and a radius E K, make an intersection at F. Also, make E a equal to half the width of the rail, and drop the perpendicular a c.

We now have all that is necessary to arrange the tangents C D, D E in a form to limit the required curve. Hence draw the horizontal line b b, Plate II, fig. 1; find the center A, then draw through A the perpendicular B B, and make A B equal to A B, figs. 1 and 2; then these lines will represent the major and minor axis of an ellipse. Now set off on each side of A the distance I F, fig. 2, and represent the points by F F; then, with F, F for centers and I M, fig. 2, for a radius, describe the curves H J H, H J H, and draw through A, and tangent to these curves, the line J J produced; then make A W equal D M, fig. 2, and erect the perpendicular W h, and with A for a center and a radius D E, fig. 2, make an intersection at C, and draw C A produced; then make A D equal C D, fig. 2. Now draw C I parallel and equal to A D, also I D parallel and equal to A C, then C I and I D are the tangents which limit the curve that will stand over A C. Now make A R equal P D, fig. 2, and the line R D, drawn perpendicular to R C, will pass through D, if the work be correct.

Now to obtain the bevels for the joints. From F let fall the perpendicular F E, and make it equal A B, or half the minor axis; then connect the intersection H and E, and at H is the bevel for the joint at C. Again, with F for a centre and a radius * touching C A produced, make an intersection at V, and connect V and E, and at V is the bevel for the joint at D. Now, to obtain the width of the mold at the joints C and D, first draw the line H n square with H E, and make it equal to half the width of the rail; then draw $n o$ square with H n, and H o is half the width of the mold at C; then draw V x square with V E, and make it half the width of the rail; also, draw $x y$ square with V x, then V y is half the width of the mold at D.

Now produce I C to 2, and make 2, 1 and 2, 3 square with 2 C, and make each equal to H o; then draw 1, 4 and 3, 5 parallel with 2 C; also set off on each side of D, square with I D, the distance V y, and we have the width of the mold at the joints 2 and D. Again, set off on each side of B, B g, and B f; these make equal to half the width of the rail, which can be found at a E, fig. 2; also, set off on each side of F F the distance $a c$, fig. 2, represented by d, c, a, b, respectively.

In the drawing, 1, 3, 5, 4 represents the shank, or portion of the straight rail, and it now remains to describe the curves through 5 $g k$ and through 4 $f m$ to complete the mold.

There are several methods to describe these, two of which we will give.

By the first method, stick a pin at d, and one at b, on the major axis; also, one at g, on the minor axis; then make a loop in the end of a linen thread, and fasten one end at d; then extend the thread around the pin at g to b, and make two or three turns around the pin at this point, and keep the finger on the end to prevent slipping; then remove the pin from g, and place a pencil, notched for the purpose, in the bight, and bring the thread drawn to 5 (see Plate IV, fig. 3, for description); then describe the curve 5 $h g k$, and remove the pins to $c f$ and a, and by a similar process describe the curve 4 $f m$.

The second method is to use a trammel, Plate IV, fig. 3, which consists of two pieces of wood of convenient length, crossing each other at right angles, each part being grooved in the center for the passage of two pins placed in a beam, as $p p$, in which is a describing pencil at t. In this case, place the center of the trammel at A, and let the arms or center of the grooves correspond with the major and mi-

---

*The radius must equal A P, fig. 2.

nor axis F A F and B A B; then find the distance with the dividers from g to b, and place one pin of the trammel beam to this distance from the describing pencil, which is half the length of the major axis; then place the other pin to the distance Ag from the pencil, this will describe the curve 5 g k; then for the inside curve make the major axis equal to the distance from f to a, and the minor axis equal to Af, this will describe the curve 4 f m.

The mold is now complete, being 3, 1 k m, with the tangents 2 C I and I D, or that part which comes within the curves marked upon it. It now remains to apply it to a piece of stuff for the purpose of forming a wreath-piece. For this purpose, procure a piece of plank of the thickness of the width of the rail; face one side, and lay the mold on, and mark the joints or ends. Also, transfer the tangent lines from the mold to the plank, then cut the ends square through, and see that the joints are square with the surface of the plank, and also with the trangents transferred from the mold; then extend the tangents across the ends, through the center of the joint, and apply the stock of bevel II to the surface at 2, and let the blade extend through the center point of the joint, and draw a line by the blade. Also, place the bevel at V, in the same position at D, and draw a line by the blade; then, from the points where these lines intersect the surface of the plank, draw the tangent lines parallel to those transferred from the mold; then lay the mold on the plank, and make the tangents on the mold correspond with those drawn by the bevel lines across the ends, and mark the surface by the edges of the mold,—the other surface must be marked by a similar process; then, by raising the wreath-piece to its place, after the edges have been worked to these lines, both edges will be found to be plumb. It now remains to bevel the top and bottom. For this purpose, set off on each side of the center point of the bevel line passing through the center, half the thickness of the rail, and from these points square lines over to the edges; this must be done at both joints. Also at B, on the minor axis, find the center of the thickness, and set off half the thickness of the rail on each side; we then have the thickness of the rail at three points on each edge, namely, at each joint, and at the minor axis; consequently, nothing more can be required to describe a mold for a wreath-piece, or the working of the same, whose center, when properly executed, will stand over A C, fig. 1.

Before leaving this example, however, it may be well to remark, that the shank, or straight wood, extends to 4, 5, and at these points the curves commence; consequently, the edges of the shank must be worked to the bevel II. Then the line, marked through the center at right angles to the bevel, must be extended along the edge to the distance 2 C, and through this point mark the plumb-line. This gives the points where the curves commence on the top and bottom of the rail.

SECOND EXAMPLE.

The last example gives the mold for the quarter circle. We will now give an example to describe a mold for a wreath-piece to any arc of a circle.

Let A C X, Plate III, fig. 1, represent a cylinder or line passing through the center of the balusters, over which the center of a wreath is required to stand. Now it is proposed to describe a mold whose center will stand over the arc A E, fig. 1. Hence there will be a joint at A, and one at E; and, as in the last example, it is necessary to know the diameter of the cylinder, the pitch of the rail, and the height of the center of the joint at E, above A. For the above purpose, the first thing in order is to draw a tangent to A E C at E; hence O E must form a right angle with J E, which is the required tangent; also, draw D E square with A B. Now A B and B C are tangents to the quarter circle, as in the last example, and may be spread out at C A B, fig. 2, parallel to A O X; next extend the line B A through C, fig. 2, indefinitely; also, extend C O to A, and M X to B, at fig. 2. We now have C A B parallel and equal to O A X; also, the perpendicular lines at C, A, and B. Now make A D equal to A D, fig. 1, and draw a line from D to L, to the same pitch as the straight rail, and produce it to P; then draw the line K Q a parallel to C A B, and to the same height above it that the center of the joint at E is above A, fig. 1. Then make D J equal A J, and J H* equal J E, fig. 1; and, also, raise the perpendiculars J N and H O, and connect N and O. Now D N and N O is the elevations of the tangents, indicating joints at D and O; hence make M O equal A B, figs. 1 and 2, and make a point. Also, make Q E equal D E, fig. 1, and draw a line from L, through E, to an intersection with B X at R; also, draw C Z R parallel with K Q a. Now produce P L to an intersection with C Z R at Y, and make H U equal Z L, and connect U and M; also, make S R equal P L, and let fall the indefinite perpendicular S t. Now draw the line U V perpendicular to N O through U; also, a perpendicular to P Y from W through Z; then with Z for a center, and a radius Z W, make an intersection at X, and draw a line from R, through X, to an intersection at T. Again, with R for a center and a radius R T, make an intersection at F, and connect F and R; then make R b equal half the width of the rail, and let fall the perpendicular b d.

We now have the parts necessary for our purpose. Hence, the next

---

* A J and J E are always equal.

thing is to draw a horizontal line, *t* J, Plate IV, fig. 1, and through the center Z draw Z B, and make it equal to A B, fig. 2. These lines again represent the major and minor axis of an ellipse. Hence, set off on each side of Z the distance Z F for the foci; then with F F for centers, and Z W, fig. 2, for a radius, describe the curves D L, D L, and draw L L through the center and tangent to these curves. Again, with F F for centers, and a radius U V, fig. 2, describe the curves H N, H N, and draw N N * through Z, and tangent to these curves also. Then make W Z equal L W, and Z V equal V O, fig. 2; then raise the perpendicular W *w* and V *v* to these lines, and, with Z for a center and a radius L R, fig. 2, make an intersection at Y, and connect Y and Z. Also, with Z for a center and a radius equal to M U, fig. 2, make an intersection at M, and draw M Z; then draw Y O square with W *w*, and parallel with W L. Also, draw M O square with V *v*, and parallel with V N. Now these lines intersect at O; and, if correct, must equal D N and N O, fig. 2,—that is, Y O must equal D N, and O M must equal N O.

To obtain the bevels, make F R equal Z B, or half the minor axis; then connect the intersection at D with R, and at D is the bevel for the shank. Also, connect H and R, and at H is the bevel for the joint M. Now find the width of the mold, as in the last example, at the joints Y and M,—D *t* being the width at Y, and H 1 the width at M. Now produce the tangent O Y to 2, and set off on each side the distance D *t* square with it, and represent the points by 1, 3; then draw 1, 4 and 3, 5 parallel to 2 Y. Also, set off on each side of M the distance H 1; next set off on each side of B the distance R *b*, fig. 2, or half the width of the rail. Also, on each side of F F, set off the distance *b d*, fig. 2, for the foci to the curves, and represent the points respectively by *o*, P, *m*, *n*. Then stick a pin at *o* and *n*, on the major axis; also at *b*, on the minor axis, and proceed to describe the curves for the mold, as in the last example.

Hence, knowing the diameter of the cylinder, the pitch of the straight rail, and the height of one joint of a wreath-piece above the other, enables the workman by the above process to describe a mold for any arc of a circle with the greatest exactness.

NOTE. When drawing perpendicular or parallel lines, greater accuracy will be obtained if the method for drawing the same, given in the problems, be followed, than will by using a square, which is sometimes practiced.

Reference was made in the description to Plate I, fig. 1, to the method of obtaining the pitch of a rail, connecting with a wreath-

---

* N N is an indefinite line, but by a blunder of the engraver it is made to intersect with the tangent Y O at 2, which is unnecessary.

piece, and also the height of one end of the same above the other. Now the elevation from which these are obtained requires a particular description; hence, let fig. 1, Plate I, be the plan as laid down on the floor. Also, let 1, 2 be on the line of the face of the last riser, connecting with the platform, or landing; then place the balusters equal distances apart, the same as on the regular steps. Hence, there will be a baluster at A, one at 2, and one at C; the one at A being on the last step, and at the point where the circle commences; while those at 2 and C are on the platform, or landing. Now there are two balusters of different lengths on each of the regular steps, the shortest being on the same line as the face of the riser; hence, one of the shortest balusters must be placed at 2, and one of the longest at C. Again, at fig. 2, let 4, 1 be the last step, 1, 2 the last riser, and 2, 3 the landing; this shows part of the elevation of the same. Now the bottom of the rail is supposed to be resting on the stairs; hence C P represents the pitch and center of the same, consequently is one-half the thickness of the rail above the corner of the stairs, as seen at 2, fig. 2, where must be placed a short ladder. Now as the next is at the joint C, and being a long one, it follows that the height of the center of the joint at C, above the landing or floor, should be equal to half the height of a riser, added to half the thickness of the rail. This is near enough, though not always practiced; hence the line W E is drawn to this height above the platform 2, 3, fig. 2. This applies to platform stairs, where two balusters, as at 2 and C, are on the platform; also, to a gallery rail; but the bevel rail connecting should be from one-half to one inch higher than the joint at C. And it should also be remembered, that the line of the last riser in the elevation is obtained by making A 5, fig. 2, equal B 1, fig. 1.

But in Plate III, fig. 1, there are winders, A 1, E and C being the equal divisions; hence the wreath-piece in passing from A to E passes over two steps, consequently rises to the height of two risers. Hence the line K Q a, fig. 2, must be drawn to the height of two risers above the line C B. Sometimes it is not convenient to place the joints directly over the risers; but this case presents no difficulty whatever, if the elevation be so drawn as to give the pitch of the rail and true position of the joints; this requires no further explanation.

THE CARPENTERS' AND BUILDERS' GUIDE. 37

BOARD, PLANK, AND SCANTLING MEASURE.

Width	1	2	3	4 2x2	5	6 2x3	7	8 2x4	9 3x3	10 2x5	11	12 2x6 3x4	13
Length. 1	.1	.2	.3	.4	.5	.6	.7	.8	.9	.10	.11	1.	1. 1
2	.2	.4	.6	.8	.10	1.	1. 2	1. 4	1. 6	1. 8	1.10	2.	2. 2
3	.3	.6	.9	1.	1. 3	1. 6	1. 9	2.	2. 3	2. 6	2. 9	3.	3. 3
3½	.3	.7	.10	1. 2	1. 5	1. 9	2.	2. 4	2. 7	2.11	3. 2	3. 6	3. 9
4	.4	.8	1.	1. 4	1. 8	2.	2. 4	2. 8	3.	3. 4	3. 8	4.	4. 4
4½	.4	.9	1. 1	1. 6	1.10	2. 3	2. 7	3.	3. 4	3. 9	4. 1	4. 6	4.10
5	.5	.10	1. 3	1. 8	2. 1	2. 6	2.11	3. 4	3. 9	4. 2	4. 7	5.	5 5
5½	.5	.11	1. 4	1.10	2. 3	2. 9	3. 2	3. 8	4. 1	4. 7	5.	5. 6	5.11
6	.6	1.	1. 6	2.	2. 6	3.	3. 6	4.	4 6	5.	5. 6	6.	6. 6
6½	.6	1. 1	1. 7	2. 2	2. 8	3. 3	3. 9	4. 4	4.10	5. 5	5.11	6. 6	7.
7	.7	1. 2	1. 9	2. 4	2.11	3. 6	4. 1	4. 8	5. 3	5.10	6. 5	7.	7. 7
7½	.7	1. 3	1.10	2 6	3. 1	3. 9	4. 4	5.	5. 7	6. 3	6.10	7. 6	8. 1
8	.8	1. 4	2.	2 8	3. 4	4.	4. 8	5. 4	6.	6. 8	7. 4	8.	8. 8
8½	.8	1. 5	2. 1	2.10	3. 6	4. 3	4.11	5. 8	6. 4	7. 1	7. 9	8. 6	9. 2
9	.9	1. 6	2. 3	3.	3. 9	4. 6	5. 3	6.	6. 9	7. 6	8. 3	9.	9. 9
9½	.9	1. 7	2. 4	3. 2	3.11	4. 9	5. 6	6. 4	7. 1	7.11	8. 8	9. 6	10. 3
10	.10	1. 8	2. 6	3. 4	4. 2	5.	5.10	6. 8	7. 6	8. 4	9. 2	10.	10.10
10½	.10	1. 9	2. 7	3. 6	4. 4	5. 3	6. 1	7.	7.10	8. 9	9. 7	10. 6	11. 4
11	.11	1.10	2. 9	3. 8	4. 7	5. 6	6. 5	7. 4	8. 3	9. 2	10. 1	11.	11.11
11½	.11	1.11	2.10	3.10	4. 9	5. 9	6. 8	7. 8	8. 7	9. 7	10. 6	11. 6	12. 5
12	1.	2.	3.	4.	5.	6.	7.	8.	9.	10.	11.	12.	13.
12½	1.	2. 1	3. 1	4. 2	5. 2	6. 3	7. 3	8. 4	9. 4	10. 5	11. 5	12. 6	13. 6
13	1. 1	2. 2	3. 3	4. 4	5. 5	6. 6	7. 7	8. 8	9. 9	10.10	11.11	13.	14. 1
13½	1. 1	2. 3	3. 4	4. 6	5. 7	6. 9	7 10	9.	10. 1	11. 3	12. 4	13. 6	14. 7
14	1. 2	2. 4	3. 6	4. 8	5.10	7.	8. 2	9. 4	10. 6	11. 8	12.10	14.	15. 2
14½	1. 2	2. 5	3. 7	4.10	6.	7. 3	8. 5	9. 8	10.10	12. 1	13. 3	14. 6	15. 8
15	1. 3	2. 6	3. 9	5.	6. 3	7. 6	8. 9	10.	11. 3	12. 6	13. 9	15.	16. 3
15½	1. 3	2. 7	3.10	5. 2	6. 5	7. 9	9.	10. 4	11. 7	12.11	14. 2	15. 6	16. 9
16	1. 4	2. 8	4.	5. 4	6. 8	8.	9. 4	10. 8	12.	13. 4	14. 8	16.	17. 4
16½	1. 4	2. 9	4. 1	5. 6	6.10	8. 3	9. 7	11.	12. 4	13. 9	15. 1	16. 6	17.10
17	1. 5	2.10	4. 3	5. 8	7. 1	8. 6	9.11	11. 4	12. 9	14. 2	15. 7	17.	18. 5
17½	1. 5	2.11	4. 4	5.10	7. 3	8. 9	10. 2	11. 8	13. 1	14. 7	16.	17. 6	18.11
18	1. 6	3.	4. 6	6.	7. 6	9.	10. 6	12.	13. 6	15.	16. 6	18.	19. 6
18½	1. 6	3. 1	4. 7	6. 2	7. 8	9. 3	10. 9	12. 4	13.10	15. 5	16.11	18. 6	20.
19	1. 7	3. 2	4. 9	6. 4	7.11	9. 6	11. 1	12. 8	14. 3	15.10	17. 5	19.	20. 7
19½	1. 7	3. 3	4.10	6. 6	8. 1	9. 9	11. 4	13.	14. 7	16. 3	17.10	19. 6	21. 1
20	1. 8	3. 4	5.	6. 8	8. 4	10.	11. 8	13. 4	15.	16. 8	18. 4	20.	21. 8
20½	1. 8	3. 5	5. 1	6.10	8. 6	10. 3	11.11	13. 8	15. 4	17. 1	18. 9	20. 6	22. 2
21	1. 9	3. 6	5. 3	7.	8. 9	10. 6	12. 3	14.	15. 9	17. 6	19. 3	21.	22. 9
21½	1. 9	3. 7	5. 4	7. 2	8.11	10. 9	12. 6	14. 4	16. 1	17.11	19. 8	21. 6	23. 3
22	1.10	3. 8	5. 6	7. 4	9. 2	11.	12.10	14. 8	16. 6	18. 4	20. 2	22.	23.10
22½	1.10	3. 9	5. 7	7. 6	9. 4	11. 3	13. 1	15.	16.10	18. 9	20. 7	22. 6	24. 4
23	1.11	3.10	5. 9	7. 8	9. 7	11. 6	13. 5	15. 4	17. 3	19. 2	21. 1	23.	24.11
23½	1.11	3.11	5.10	7.10	9. 9	11. 9	13. 8	15. 8	17. 7	19. 7	21. 6	23. 6	25. 5
24	2.	4.	6.	8.	10.	12.	14.	16.	18.	20.	22.	24.	26.
24½	2.	4. 1	6. 1	8. 2	10. 2	12. 3	14. 3	16. 4	18. 4	20 5	22. 5	24. 6	26. 6
25	2. 1	4. 2	6. 3	8. 4	10. 5	12. 6	14. 7	16. 8	18. 9	20.10	22.11	25.	27. 1
25½	2. 1	4. 3	6. 4	8. 6	10. 7	12. 9	14.10	17.	19. 1	21. 3	23. 4	25. 6	27. 7
26	2. 2	4. 4	6. 6	8. 8	10.10	13.	15. 2	17. 4	19. 6	21. 8	23.10	26.	28. 2
26½	2. 2	4. 5	6. 7	8.10	11.	13. 3	15. 5	17. 8	19.10	22. 1	24. 3	26. 6	28. 8
27	2. 3	4. 6	6. 9	9.	11. 3	13. 6	15. 9	18.	20. 3	22. 6	24. 9	27.	29. 3
27½	2. 3	4. 7	6.10	9. 2	11. 5	13. 9	16.	18. 4	20. 7	22 11	25. 2	27. 6	29. 9
28	2. 4	4. 8	7.	9. 4	11. 8	14.	16. 4	18. 8	21.	23. 4	25. 8	28.	30. 4
28½	2. 4	4. 9	7. 1	9. 6	11.10	15. 3	16. 7	19.	21. 4	23. 9	26. 1	28. 6	30.10
29	2. 5	4.10	7. 3	9. 8	12. 1	15. 6	16.11	19. 4	21. 9	24. 2	26. 7	29.	31. 5
29½	2. 5	4.11	7. 4	9.10	12. 3	15. 9	17. 2	19. 8	22. 1	24. 7	27.	29. 6	31.11
30	2. 6	5.	7. 6	10.	12. 6	15.	17. 6	20.	22. 6	25.	27. 6	30.	32. 6

BOARD, PLANK, AND SCANTLING MEASURE—Continued.

Width	14 2x7	15 3x5	16 2x8 4x4	17	18 2x9 3x6	19	20 2x10 4x5	21 3x7	22 2x11	23	24 2x12 4x6 3x8	25 5x5	26 2x13
Length. 1	1. 2	1. 3	1. 4	1. 5	1.6	1. 7	1. 8	1. 9	1.10	1.11	2.	2. 1	2. 2
2	2. 4	2. 6	2. 8	2.10	3.	3. 2	3. 4	3. 6	3. 8	3.10	4.	4. 2	4. 4
3	3. 6	3. 9	4.	4. 3	4.6	4. 9	5.	5. 3	5. 6	5. 9	6.	6. 3	6. 6
3½	4. 1	4. 4	4. 8	4.11	5.3	5. 6	5.10	6. 1	6. 5	6. 8	7.	7. 3	7. 7
4	4. 8	5.	5. 4	5. 8	6.	6. 4	6. 8	7.	7. 4	7. 8	8.	8. 4	8. 8
4½	5. 3	5. 7	6.	6. 4	6.9	7. 1	7. 6	7.10	8. 3	8. 7	9.	9. 4	9. 9
5	5.10	6. 3	6. 8	7. 1	7.6	7.11	8. 4	8. 9	9. 2	9. 7	10.	10. 5	10.10
5½	6. 5	6.10	7. 4	7. 9	8.3	8. 8	9. 2	9. 7	10. 1	10. 6	11.	11. 5	11.11
6	7.	7. 6	8.	8. 6	9.	9. 6	10.	10. 6	11.	11. 6	12.	12. 6	13.
6½	7. 7	8. 1	8. 8	9. 2	9.9	10. 3	10.10	11. 4	11.11	12. 5	13.	13. 6	14. 1
7	8. 2	8. 9	9. 4	9.11	10.6	11. 1	11. 8	12. 3	12.10	13. 5	14.	14. 7	15. 2
7½	8. 9	9. 4	10.	10. 7	11.3	11.10	12. 6	13. 1	13. 9	14. 4	15.	15. 7	16. 3
8	9. 4	10.	10. 8	11. 4	12.	12. 8	13. 4	14.	14. 8	15. 4	16.	16. 8	17. 4
8½	9.11	10. 7	11. 4	12.	12.9	13. 5	14. 2	14.10	15. 7	16. 3	17.	17. 8	18. 5
9	10. 6	11. 3	12.	12. 9	13.6	14. 3	15.	15. 9	16. 6	17. 3	18.	18. 9	19. 6
9½	11. 1	11.10	12. 8	13. 5	14.3	15.	15.10	16. 7	17. 5	18. 2	19.	19. 9	20. 7
10	11. 8	12. 6	13. 4	14. 2	15.	15.10	16. 8	17. 6	18. 4	19. 2	20.	20.10	21. 8
10½	12. 3	13. 1	14.	14.10	15.9	16. 7	17. 6	18. 4	19. 3	20. 1	21.	21.10	22. 9
11	12.10	13. 9	14. 8	15. 7	16.6	17. 5	18. 4	19. 3	20. 2	21. 1	22.	22.11	23.10
11½	13. 5	14. 4	15. 4	16. 3	17.3	18. 2	19. 2	20. 1	21. 1	22.	23.	23.11	24.11
12	14.	15.	16.	17.	18.	19.	20.	21.	22.	23.	24.	25.	26.
12½	14. 7	15. 7	16. 8	17. 8	18.9	19. 9	20.10	21.10	22.11	23.11	25.	26.	27. 1
13	15. 2	16. 3	17. 4	18. 5	19.6	20. 7	21. 8	22. 9	23.10	24.11	26.	27. 1	28. 2
13½	15. 9	16.10	18.	19. 1	20.3	21. 4	22. 6	23. 7	24. 9	25.10	27.	28. 1	29. 3
14	16. 4	17. 6	18. 8	19.10	21.	22. 2	23. 4	24. 6	25. 8	26.10	28.	29. 2	30. 4
14½	16.11	18. 1	19. 4	20. 6	21.9	22.11	24. 2	25. 4	26. 7	27. 9	29.	30. 2	31. 5
15	17. 6	18. 9	20.	21. 3	22.6	23. 9	25.	26. 3	27. 6	28. 9	30.	31. 3	32. 6
15½	18. 1	19. 4	20. 8	21.11	23.3	24. 6	25.10	27. 1	28. 5	29. 8	31.	32. 3	33. 7
16	18. 8	20.	21. 4	22. 8	24.	25. 4	26. 8	28.	29. 4	30. 8	32.	33. 4	34. 8
16½	19. 3	20. 7	22.	23. 4	24.9	26. 1	27. 6	28.10	30. 3	31. 7	33.	34. 4	35. 9
17	19.10	21. 3	22. 8	24. 1	25.6	26.11	28. 4	29. 9	31. 2	32. 7	34.	35. 5	36.10
17½	20. 5	21.10	23. 4	24. 9	26.3	27. 8	29. 2	30. 7	32. 1	33. 6	35.	36. 5	37.11
18	21.	22. 6	24.	25. 6	27.	28. 6	30.	31. 6	33.	34. 6	36.	37. 6	39.
18½	21. 7	23. 1	24. 8	26. 2	27.9	29. 3	30.10	32. 4	33.11	35. 5	37.	38. 6	40. 1
19	22. 2	23. 9	25. 4	26.11	28.6	30. 1	31. 8	33. 3	34.10	36. 5	38.	39. 7	41. 2
19½	22. 9	24. 4	26.	27. 7	29.3	30.10	32. 6	34. 1	35. 9	37. 4	39.	40. 7	42. 3
20	23. 4	25.	26. 8	28. 4	30.	31. 8	33. 4	35.	36. 8	38. 4	40.	41. 8	43. 4
20½	23.11	25. 7	27. 4	29.	30.9	32. 5	34. 2	35.10	37. 7	39. 3	41.	42. 8	44. 5
21	24. 6	26. 3	28.	29. 9	31.6	33. 3	35.	36. 9	38. 6	40. 3	42.	43. 9	45. 6
21½	25. 1	26.10	28. 8	30. 5	32.3	34.	35.10	37. 7	39. 5	41. 2	43.	44. 9	46. 7
22	25. 8	27. 6	29. 4	31. 2	33.	34.10	36. 8	38. 6	40. 4	42. 2	44.	45.10	47. 8
22½	26. 3	28. 1	30.	31.10	33.9	35. 7	37. 6	39. 4	41. 3	43. 1	45.	46.10	48. 9
23	26.10	28. 9	30. 8	32. 7	34.6	36. 5	38. 4	40. 3	42. 2	44. 1	46.	47.11	49.10
23½	27. 5	29. 4	31. 4	33. 3	35.3	37. 2	39. 2	41. 1	43. 1	45.	47.	48.11	50.11
24	28.	30.	32.	34.	36.	38.	40.	42.	44.	46.	48.	50.	52.
24½	28. 7	30. 7	32. 8	34. 8	36.9	38. 9	40.10	42 10	44.11	46 11	49.	51.	53. 1
25	29. 2	31. 3	33. 4	35. 5	37.6	39. 7	41. 8	43. 9	45.10	47 11	50.	52. 1	54. 2
25½	29. 9	31.10	34.	36. 1	38.3	40. 4	42. 6	44. 7	46. 9	48.10	51.	53. 1	55. 3
26	30. 4	32. 6	34. 8	36.10	39.	41. 2	43. 4	45. 6	47. 8	49.10	52.	54. 2	56. 4
26½	30.11	33. 1	35. 4	37. 6	39.9	41.11	44. 2	46. 4	48. 7	50. 9	53.	55. 2	57. 5
27	31. 6	33. 9	36.	38. 3	40.6	42. 9	45.	47. 3	49. 6	51. 9	54.	56. 3	58. 6
27½	32. 1	34. 4	36. 8	38.11	41.3	43. 6	45.10	48. 1	50. 5	52. 8	55.	57. 3	59. 7
28	32. 8	35.	37. 4	39. 8	42.	44. 4	46. 8	49.	51. 4	53. 8	56.	58. 4	60. 8
28½	33. 3	35. 7	38.	40. 4	42.9	45. 1	47. 6	49.10	52. 3	54. 7	57.	59. 4	61. 9
29	33.10	36. 3	38. 8	41. 1	43.6	45.11	48. 4	50. 9	53. 2	55. 7	58.	60. 5	62.10
29½	34. 5	36.10	39. 4	41. 9	44.3	46. 8	49. 2	51. 7	54. 1	56. 6	59.	61. 5	63.11
30	35.	37. 6	40.	42. 6	45.	47. 6	50.	52. 6	55.	57. 6	60.	61. 6	65.

THE CARPENTERS' AND BUILDERS' GUIDE. 39

TIMBER MEASURE.

Width Length	3x9	3x10 5x6	3x11	3x12 4x9 6x6 2x18	4x7	4x8	4x10 5x8	4x11	4x12 6x8	5x7	5x9	5x10
1	2. 3	2. 6	2. 9	3.	2. 4	2. 8	3. 4	3. 8	4.	2.11	3. 9	4. 2
2	4. 6	5.	5. 6	6.	4. 8	5. 4	6. 8	7. 4	8.	5.10	7. 6	8. 4
3	6. 9	7. 6	8. 3	9.	7.	8.	10.	11.	12.	8. 9	11. 3	12. 6
3½	7.10	8. 9	9. 7	10. 6	8. 2	9. 4	11. 8	12.10	14.	0. 2	13. 1	14. 7
4	9.	10.	11.	12.	9. 4	10. 8	13. 4	14. 8	16.	11. 8	15.	16. 8
4½	10. 1	11. 3	12. 4	13. 6	10. 6	12.	15.	16 6	18.	13. 1	16.10	18. 9
5	11. 3	12. 6	13. 9	15.	11. 8	13. 4	16. 8	18. 4	20.	14. 7	18. 9	20.10
5½	12. 4	13. 9	15. 1	16. 6	12.10	14. 8	18. 4	20. 2	22.	16.	20. 7	22.11
6	13. 6	15.	16. 6	18.	14.	16.	20.	22.	24.	17. 6	22. 6	25.
6½	14. 7	16. 3	17.10	19. 6	15. 2	17. 4	21. 8	23.10	26.	18.11	24. 4	27. 1
7	15. 9	17. 6	19. 3	21.	16. 4	18. 8	23. 4	25. 8	28.	20. 5	26. 3	29. 2
7½	16.10	18. 9	20. 7	22. 6	17. 6	20.	25.	27. 6	30.	21.10	28. 1	31. 3
8	18.	20.	22.	24.	18. 8	21. 4	26. 8	29. 4	32.	23. 4	30.	33. 4
8½	19. 1	21. 3	23. 4	25. 6	19.10	22. 8	28. 4	31. 2	34.	24. 9	31.10	35. 5
9	20. 3	22. 6	24. 9	27.	21.	24.	30.	33.	36.	26. 3	33. 9	37. 6
9½	21. 4	23. 9	26. 1	28. 6	22. 2	25. 4	31. 8	34.10	38.	27. 8	35. 7	39. 7
10	22. 6	25.	27. 6	30.	23. 4	26. 8	33. 4	36. 8	40.	29. 2	37. 6	41. 8
10½	23. 7	26. 3	28.10	31. 6	24. 6	28.	35.	38. 6	42.	30. 7	39. 4	43. 9
11	24. 9	27. 6	30. 3	33.	25. 8	29. 4	36. 8	40. 4	44.	32. 1	41. 3	45.10
11½	25.10	28. 9	31. 7	34. 6	26.10	30. 8	38. 4	42. 2	46.	33. 6	43. 1	47.11
12	27.	30.	33.	36.	28.	32.	40.	44.	48.	35.	45.	50.
12½	28. 1	31 3	34. 4	37. 6	29. 2	33. 4	41. 8	45.10	50.	36. 5	46.10	52. 1
13	29. 3	32. 6	35. 9	39.	30. 4	34. 8	43. 4	47. 8	52.	37.11	48. 9	54. 2
13½	30. 4	33. 9	37. 1	40. 6	31. 6	36.	45.	49. 6	54.	39. 4	50. 7	56. 3
14	31. 6	35.	38. 6	42.	32. 8	37. 4	46. 8	51. 4	56.	40.10	52. 6	58. 4
14½	32. 7	36. 3	39.10	43. 6	33.10	38. 8	48. 4	53. 2	58.	42. 3	54. 4	60. 5
15	33. 9	37. 6	41. 3	45.	35.	40.	50.	55.	60.	43. 9	56. 3	62. 6
15½	34.10	38. 9	42. 7	46. 6	36. 2	41. 4	51. 8	56.10	62.	45. 2	58. 1	64. 7
16	36.	40.	44.	48.	37. 4	42. 8	53 4	58. 8	64.	46. 8	60.	66. 8
16½	37. 1	41. 3	45. 4	49. 6	38. 6	44	55.	60. 6	66.	48. 1	61.10	68. 9
17	38. 3	42. 6	46. 9	51.	39. 8	45. 4	56 8	62. 4	68.	49. 7	63. 9	70.10
17½	39. 4	43	48. 1	52. 6	40.10	46. 8	58. 4	64. 2	70.	51.	65. 7	72.11
18	40. 6	45.	49. 6	54.	42.	48.	60.	66.	72.	52. 6	67. 6	75.
18½	41. 7	46. 3	50.10	55. 6	43. 2	49. 4	61. 8	67.10	74.	53.11	69. 4	77. 1
19	42. 9	47. 6	52. 3	57.	44. 4	50. 8	63. 4	69. 8	76.	55. 5	71. 3	79. 2
19½	43.10	48. 9	53. 7	58. 6	45. 6	52.	65.	71. 6	78.	56.10	73. 1	81. 3
20	45.	50.	55.	60.	46 8	53. 4	66. 8	73. 4	80.	58. 4	75.	83. 4
20½	46. 1	51. 3	56. 4	61. 6	47.10	54. 8	68. 4	75. 2	82.	59. 9	76.10	85. 5
21	47. 3	52. 6	57. 9	63.	49.	56.	70.	77.	84.	61. 3	78. 9	87. 6
21½	48. 4	53. 9	59. 1	64. 6	50. 2	57. 4	71. 8	78.10	86.	62. 8	80. 7	89. 7
22	49. 6	55.	60. 6	66.	51. 4	58. 8	73. 4	80. 8	88.	64. 2	82. 6	91. 8
22½	50. 7	56. 3	61.10	67. 6	52. 6	60.	75.	82. 6	90.	65. 7	84. 4	93. 9
23	51. 9	57. 6	63. 3	69.	53. 8	61. 4	76. 8	84. 4	92.	67. 1	86. 3	95.10
23½	52.10	58. 9	64. 7	70. 6	54.10	62. 8	78. 4	86. 2	94.	68. 6	88. 1	97.11
24	54.	60.	66.	72.	56.	64.	80.	88.	96.	70.	90.	100.
24½	55. 1	61. 3	67. 4	73. 6	57. 2	65. 4	81. 8	89.10	98.	71. 5	91.10	102. 1
25	56. 3	62. 6	68. 9	75.	58. 4	66. 8	83. 4	91. 8	100.	72.11	93. 9	104. 2
25½	57. 4	63. 9	70. 1	76. 6	59. 6	68.	85.	93. 6	102.	73. 4	95. 7	106. 3
26	58. 6	65.	71. 6	78.	60. 8	69. 4	86. 8	95. 4	104.	75.10	97. 6	108. 4
26½	59. 7	66. 3	72.10	79. 6	61.10	70. 8	88. 4	97. 2	106.	76. 3	99. 4	110. 5
27	60. 9	67. 6	74. 3	81.	63.	72.	90.	99.	108.	78. 9	101. 3	112. 6
27½	61.10	68. 9	75. 7	82. 6	64. 2	73. 4	91. 8	100.10	110.	80. 2	103. 1	114. 7
28	63.	70.	77.	84.	65. 4	74. 8	93. 4	102. 8	112.	81. 8	105.	116. 8
28½	64. 1	71. 3	78. 4	85. 6	66. 6	76.	95.	104. 6	114.	83. 1	106.10	118. 9
29	65. 3	72. 6	79. 9	87.	67. 8	77. 4	96. 8	106. 4	116.	84. 7	108. 9	120.10
29½	66. 4	73. 9	81. 1	88. 6	68.10	78. 8	98. 4	108. 2	118.	86.	110. 7	122.11
30	67. 6	75.	82. 6	90.	70.	80.	100.	110.	120.	87. 6	112. 6	125.

TIMBER MEASURE—Continued.

Width	5x11	5x12 6x10	6x7	6x9	6x11	6x12	7x7	7x8	7x9	7x10	7x11	7x12
Length 1	4. 7	5.	3.6	4.6	5.6	6.	4. 1	4.8	5. 3	5.10	6. 5	7
2	9. 2	10.	7.	9.	11.	12.	8. 2	9.4	10. 6	11. 8	12.10	14
3	13. 9	15.	10.6	13.6	16.6	18.	12. 3	14.	15. 9	17. 6	19. 3	21.
3½	16.	17.6	12.3	15.9	19.3	21.	14⁻ 3	16.4	18. 4	20. 5	22. 5	24.6
4	18. 4	20.	14.	18.	22.	24.	16. 4	18.8	21.	23. 4	25. 8	28.
4½	20. 7	22.6	15.9	20.3	24.9	27.	18. 4	21.	23. 7	26. 3	28.10	31.6
5	22.11	25.	17 6	22.6	27.6	30.	20. 5	23.4	26. 3	29. 2	32. 1	35.
5½	25. 2	27.6	19.3	24.9	30.3	33.	22. 5	25.8	28.10	32. 1	35. 3	38 6
6	27. 6	30.	21.	27.	33.	36.	24. 6	28.	31. 6	35.	38. 6	42.
6½	29. 9	32.6	22.9	29.3	35.9	39.	26. 6	30.4	34. 1	37.11	41. 8	45.6
7	32. 1	35.	24.6	31.6	38.6	42.	28. 7	32 8	36. 9	40.10	44.11	49.
7½	34. 4	37.6	26.3	33.9	41.3	45.	30. 7	35.	39. 4	43. 9	48. 1	52.6
8	36. 8	40.	28.	36.	44.	48.	32. 8	37.4	42.	46. 8	51. 4	56.
8½	38.11	42.6	29.9	38.3	46.9	51·	34. 8	39.8	44. 7	49. 7	54. 6	59.6
9	41. 3	45.	31.6	40.6	49.6	54.	36. 9	42.	47. 3	52. 6	57. 9	63.
9½	43. 6	47.6	33.3	42.9	52.3	57.	38. 9	44.4	49.10	55. 5	60.11	66.6
10	45.10	50.	35.	45.	55.	60.	40.10	46.8	52. 6	58. 4	64. 2	70.
10½	48. 1	52.6	36.9	47.3	57.9	63.	42.10	49.	55. 1	61. 3	67. 4	73.6
11	50. 5	55.	38.6	49.6	60.6	66.	44.11	51.4	57. 9	64. 2	70. 7	77.
11½	52. 8	57.6	40.3	51.9	63.3	69.	46.11	53.8	60. 4	67. 1	73. 9	80.6
12	55.	60.	42.	54.	66.	72.	49.	56.	63.	70.	77.	84.
12½	57. 3	62.6	43.9	56.3	68 9	75.	51.	58.4	65. 7	72.11	80. 2	87.6
13	59. 7	65.	45.6	58.6	71.6	78.	53. 1	60.8	68. 3	75.10	83. 5	91.
13½	61.10	67.6	47.3	60.9	74.3	81.	55. 1	63.	70.10	78. 9	86. 7	94.6
14	64. 2	70.	49.	63.	77.	84.	57. 2	65.4	73. 6	81. 8	89.10	98.
14½	66. 5	72.6	50.9	65.3	79.9	87.	59. 2	67.8	76. 1	84. 7	93.	101.6
15	68. 9	75.	52.6	67.6	82.6	90.	61. 3	70.	78. 9	87. 6	96. 3	105.
15½	71.	77.6	54.3	69.9	85.3	93.	63. 3	72.4	81. 4	90. 5	99. 5	108.6
16	73. 4	80.	56.	72.	88.	96.	65. 4	74.8	84.	93. 4	102. 8	112.
16½	75. 7	82.6	57.9	74.3	90.9	99.	67. 4	79.	86. 7	96. 3	105.10	115.6
17	77.11	85.	59.6	76.6	93.6	102.	69. 5	79.4	89. 3	99. 2	109. 1	119.
17½	80. 2	87.6	61.3	78.9	96.3	105.	71. 5	81.8	91.10	102. 1	112. 3	122.6
18	82. 6	90.	63.	81.	99.	108.	73. 6	84.	94. 6	105.	115. 6	126.
18½	84. 9	92.6	64.9	83.3	101.9	111.	75. 6	86.4	97. 1	107.11	118. 8	129.6
19	87. 1	95.	66.6	85.6	104.6	114.	77. 7	88.8	99. 9	110.10	121.11	133.
19½	89 4	97.6	68.3	87.9	107.3	117.	79. 7	91.	102. 4	113. 9	125. 1	136.6
20	91. 8	100.	70.	90.	110.	120.	81. 8	93.4	105.	116. 8	128. 4	140.
20½	93.11	102.6	71 9	92.3	112.9	123.	83. 8	95.8	107. 7	119. 7	131. 6	143.6
21	96. 3	105.	73.6	94.6	115.6	126.	85. 9	98.	110. 3	122. 6	134. 9	147.
21½	98. 6	107.6	75.3	96.9	118.3	129.	87. 9	100.4	112.10	125. 5	137.11	150.6
22	100.10	110.	77.	99.	121.	132.	89 10	102.8	115. 6	128. 4	141. 2	154.
22½	103. 1	112 6	78.9	101.3	123.9	135.	91.10	105.	118. 1	131. 3	144. 4	157.6
23	105. 5	115.	80.6	103.6	126.6	138.	93.11	107.4	120. 9	134. 2	147. 7	161.
23½	107. 8	117.6	82.3	105.9	129.3	141.	95.11	109.8	123. 4	137. 1	150. 9	164.6
24	110.	120.	84.	108.	132.	144.	98.	112.	126.	140.	154.	168.
24½	112. 3	122.6	85.9	110.3	134.9	147.	100.	114.4	128. 7	142.11	157. 2	171.6
25	114. 7	125.	87.6	112.6	137.6	150.	102. 1	116.8	131. 3	145.10	160. 5	175.
25½	116.10	127.6	89.3	114.9	140.3	153.	104. 1	119.	133.10	148. 9	163. 7	178.6
26	119. 2	130.	91.	117.	143.	156.	106. 2	121.4	136. 6	151. 8	166.10	182.
26½	121. 5	132.6	92.9	119.3	145.9	159.	108. 2	123.8	138. 1	154. 7	170.	185.6
27	123. 9	135.	94.6	121.6	148.6	162.	110. 3	126.	141. 9	157. 6	173. 3	189.
27½	126.	137.6	96.3	123.9	151.3	165.	112. 3	128.4	144. 4	160. 5	176. 5	192.6
28	128. 4	140.	98.	126.	154.	168.	114. 4	130.8	147.	163. 4	179. 8	196.
28½	130. 7	142.6	99.9	128.3	156.9	171·	116. 4	133.	149. 7	166. 3	182.10	199.6
29	132.11	145.	101.6	130.6	159.6	174.	118. 5	135.4	152. 3	169. 2	186. 1	203.
29½	135. 2	147.6	103.3	133.9	162.3	177.	120. 5	137.8	154.10	172. 1	189. 3	206.6
30	137. 6	150.	105.	135.	165.	180.	122. 6	140.0	157. 6	175.	191. 6	210.

TIMBER MEASURE—Continued.

Width / Length	8x8	8x10	8x11	8x12	9x9	9x10	9x11	9x12	10x10	10x11	10x12	11x11
1	5.4	6.8	7.4	8.	6. 9	7.6	8. 3	9.	8. 4	9. 2	10.	10. 1
2	10.8	13.4	14.8	16.	13. 6	15.	16. 6	18.	16. 8	18. 4	20.	20. 2
3	16.	20.	22.	24.	20. 3	22.6	24. 9	27.	25.	27. 6	30.	30. 3
3½	18.8	23.4	25.8	28.	23. 7	26.3	28.10	31.6	29. 2	32. 1	35.	35. 3
4	21.4	26.8	29.4	32.	27.	30.	33.	36.	33. 4	36. 8	40.	40. 4
4½	24.	30.	33.	36.	30. 4	33.9	37. 1	40.6	37. 6	41. 3	45.	45. 4
5	26.8	33.4	36.8	40.	33. 9	37.6	41. 3	45.	41. 8	45.10	50.	50. 5
5½	29.4	36.8	40.4	44.	37. 1	41.3	45. 4	49.6	45.10	50. 5	55.	55. 5
6	32.	40.	44.	48.	40. 6	45.	49. 6	54.	50.	55.	60.	60. 6
6½	34.8	43 4	47.8	52.	43.10	48.9	53. 7	58.6	54. 2	59. 7	65.	65. 6
7	37.4	46.8	51.4	56.	47. 3	52.6	57. 9	63.	58. 4	64. 2	70.	70. 7
7½	40.	50.	55.	60.	50. 7	56.3	61.10	67.6	62. 6	68. 9	75.	75. 7
8	42.8	53.4	58.8	64.	54.	60.	66.	72.	66. 8	73. 4	80.	80. 8
8½	45.4	56.8	62.4	68.	57. 4	63.9	70. 1	76 6	70.10	77.11	85.	85. 8
9	48.	60.	66.	72.	60. 9	67.6	74. 3	81.	75.	82. 6	90.	90. 9
9½	50.8	63.4	69.8	76.	64. 1	71.3	78. 4	85.6	79. 2	87. 1	95.	95. 9
10	53.4	66.8	73.4	80.	67. 6	75.	82. 6	90.	83. 4	91. 8	100.	100.10
10½	56.	70.	77.	84.	70.10	78.9	86. 7	94.6	87. 6	96. 3	105.	105.10
11	58.8	73.4	80.8	88.	74. 3	82.6	90. 9	99.	91. 8	100.10	110.	110.11
11½	61.4	76.8	84.4	92.	77. 7	86.3	94.10	103.6	95 10	105. 5	115.	115.11
12	64.	80.	88.	96.	81.	90.	99.	108.	100.	110.	120.	121.
12½	66.8	83.4	91.8	100.	84. 4	93.9	103. 1	112.6	104. 2	114. 7	125.	126.
13	69.4	86.8	95.4	104.	87. 9	97.6	107. 3	117.	108. 4	119. 2	130.	131. 1
13½	72.	90.	99.	108.	91. 1	101.3	111. 4	121.6	112. 6	123. 9	135.	136. 1
14	74.8	93.4	102.8	112.	94. 6	105.	115. 6	126.	116. 8	128. 4	140.	141. 2
14½	77.4	96.8	106.4	116.	97.10	108.9	119. 7	130.6	120.10	132.11	145.	146. 2
15	80.	100.	110.	120.	101. 3	112.6	123. 9	135.	125.	137. 6	150.	151. 3
15½	82.8	103.4	113.8	124.	104. 7	116.3	127.10	139.6	129. 2	142. 1	155.	156. 3
16	85.4	106.8	117.4	128.	108.	120.	132.	144.	133. 4	146. 8	160.	161. 4
16½	88.	110.	121.	132.	111. 4	123.9	136. 1	148.6	137. 6	151. 3	165.	166. 4
17	90.8	113.4	124.8	136.	114. 9	127.6	140. 3	153.	141. 8	154.10	170.	171. 5
17½	93.4	116.8	128.4	140.	118. 1	131.3	144. 4	157.6	145.10	159. 5	175.	176. 5
18	96.	120.	132.	144.	121. 6	135.	148. 6	162.	150.	165.	180.	181. 6
18½	98.8	123.4	135.8	148.	124.10	138.9	152. 7	166.6	154. 2	169. 7	185.	186. 6
19	101.4	126.8	139.4	152.	128. 3	142.6	156. 9	171.	158. 4	174. 2	190.	191. 7
19½	104.	130.	143.	156.	131. 7	146 3	160.10	175.6	162. 6	178. 9	195.	196. 7
20	106.8	133.4	146.8	160.	135.	150.	165.	180.	166. 8	183. 4	200.	201. 8
20½	109.4	136.8	150.4	164.	138. 4	153.9	169. 1	184.6	170.10	187.11	205.	206. 8
21	112.	140.	154.	168.	141. 9	157.6	173. 3	189.	175.	192. 6	210.	211. 9
21½	114.8	143.4	157.8	172.	145. 1	161.3	177. 4	193.6	179. 2	197. 1	215.	216. 9
22	117.4	146.8	161 4	176.	148. 6	165.	181. 6	198.	183. 4	201. 8	220.	221.10
22½	120.	150.	165.	180.	151.10	168.9	185. 7	202.6	187. 6	206. 3	225.	226.10
23	122.8	153.4	168.8	184.	155. 3	172.6	189. 9	207.	191. 8	210.10	230.	231.11
23½	125.4	156.8	172.4	188.	158. 7	176.3	193.10	211.6	195.10	215. 5	235.	236.11
24	128.	160.	176.	192.	162.	180.	198.	216.	200.	220.	240.	242.
24½	130.8	163.4	179.8	196.	165. 4	183.9	202. 1	220.6	204. 2	224. 7	245.	247.
25	133.4	166.8	183.4	200.	168. 9	187.6	206. 3	225.	208. 4	229. 2	250.	252. 1
25½	136.	170.	187.	204.	172. 1	191.3	210. 4	229.6	212. 6	233. 9	255.	257. 1
26	138.8	173.4	190.8	208.	175. 6	195.	214. 6	234.	216. 8	238. 4	260.	262. 2
26½	141.4	176.8	194.4	212.	178.10	198.9	218. 7	238.6	220.10	242.11	265	267. 2
27	144.	180.	198.	216.	182. 3	202.6	222. 9	243.	225.	247. 6	270.	272. 3
27½	146.8	183.4	201.8	220.	185. 7	206.3	226.10	247.6	229. 2	252. 1	275.	277. 3
28	149.4	186.8	205.4	224.	189.	210.	231.	252.	233. 4	256. 8	280.	282. 4
28½	152.	190.	209.	228.	192. 4	213.9	235. 1	256.6	237. 6	261. 3	285.	287. 4
29	154.8	193.4	212.8	232.	195. 9	217.6	239. 3	261.	241. 8	265.10	290.	292. 5
29½	157.4	196.8	216.4	236.	199. 1	221.3	243. 4	265.6	245.10	270. 5	295.	297. 5
30	160.	200.	220.	240	202. 6	225.	247. 6	270.	250.	275.	300.	302. 6

FORM OF BUILDER'S CONTRACT.

FROM "PARSONS' LAW OF BUSINESS," BY PERMISSION OF THE PUBLISHERS.

An Agreement, of two parts, made this   day of
            in the year one thousand eight hundred and sixty-
by and between   part of the first part, and

part of the second part.
 The said part of the first part, in consideration of the sum of money to be paid by the said part of the second part, as hereinafter mentioned, and the covenants and agreements hereinafter recited, to be kept and performed by the said part of the second part, do
for   sel   and   **Covenant, Promise,** and
Agree, to and with the said part of the second part,
that   the said part of the first part, shall and will, in a good and workmanlike manner, and according to the best of art and ability, do and perform the following work, and provide
            materials for the same, that is to say:

 The whole of said work is to be performed, and all the said materials furnished, in conformity with the plans and specifications of the same, as made by   the
ARCHITECT hereby appointed by said part of the second part, which plans and specifications bear even date herewith, and are signed by the parties hereto, and under the superintendence and direction of
hereby appointed SUPERINTENDENT and AGENT of the said part of the second part, which plans and specifications are to be considered as forming a part of this agreement, as if herein fully written and drawn.
 The said part of the first part further agree that the work aforesaid shall be commenced and be constantly prosecuted, and the materials aforesaid promptly furnished   and that all said work shall be completed on or before the   day of
in the year one thousand eight hundred and sixty   and, furthermore, that no charge of any kind shall be made by the said part of the first part to the said part of the second part, beyond the **sum of**

dollars, unless the said part   of the second part, and the said Superintendents, shall alter the aforesaid plans and specifications, in which case the value of such alterations shall be added to the amount to be paid under this contract, or deducted therefrom, as the case may require: it being expressly understood that no extra work of any kind shall be performed, or extra materials furnished, by the said part   of the first part, unless first authorized by the said part   of the second p rt, and the said Superintendents, in writing; and that the said part of the second part, and the said Superintendents may, from time to time, make any alterations of, to, and in the said plans and specifications, upon the terms aforesaid.

The said part   of the first part, for   sel   and   legal representatives, further promise   and agree   that insurance shall be effected upon the building as soon as the roof is put on and covered; the amount of said insurance to be for such sum as the said part   of the second part, and the said Superintendents shall direct, to be further increased, from time to time, at the direction of the said party of the second part, and the said Superintendents; the policy to be in the name and for the benefit of said part   of the second part, or legal representatives, and to be made payable, in case of loss, to   for whom it may concern:—each party to this agreement hereby agreeing to pay one-half the cost of such insurance.

The said part   of the second part, for   sel   and   legal representatives, in consideration of the materials being provided and the labor done as herein required, and all other of the stipulations, requirements, matters and things herein set forth, being kept and performed by said part   of the first part, **Covenant, Promise,** and **Agree,** to and with the said part   of the first part: that   will well and truly pay, or cause to be paid, unto the said part of the first part, or   legal representatives, the sum of   dollars, in the manner following :—

It is agreed by and between the parties to this agreement, as follows:—

1st. That for each and every day's delay in the performance and completion of this agreement, or of any extra work under it, after the said   day of   in the year one thousand eight hundred and   there shall be allowed and paid by said part   of the first part, to said part   of the second part, or   representatives, damages for such delay, if the same shall arise from any act or default on the part of the said part   of the first part

2d. That the said part of the first part, or representatives, shall not be delayed in the constant progress of the work under this agreement, or any of the extra work under the same or connected therewith, by said party of the second part, or by his Superintendents or any other contractor employed by the said part of the second part, upon or about the premises; and for each and every day, if any, shall be so delayed, additional day to be allowed to complete the work aforesaid, from and after the day hereinbefore appointed for its entire completion, unless upon the contingency provided for below in the 5th article.

3d. That each and every person employed, by sub-contract or "piece work," by the said part of the first part, in the providing materials or performing labor or works in the fulfilment or execution of this agreement, shall be, in the opinion of the said Superintendents, a suitable, competent, and satisfactory person.

4th. That the said part of the first part shall and will engage and provide, at own cost and expense, during the progress of the works under, and until the completion and fulfilment of this agreement, a thoroughly competent "Foreman of the Works," whose duty it shall be to attend to the general supervision of all matters hereby undertaken by said part of the first part, and also to the correct and exact making, preparing, laying-out, and locating of all patterns, molds, models, and measurements in, to, for, and upon the works hereby agreed upon, from and in conformity with the said plans and specifications, and according to the direction of said Architects.

5th. That if at any time during the progress of the work the said Superintendents shall find that said work is not carried forward with sufficient rapidity and thoroughness, or that the materials furnished, foreman of the works, sub-contractors or workmen employed by the part of the first part, are unsatisfactory, and insufficient for the completion of the work within the time and in the manner stipulated in the plans and specifications aforesaid, shall give notice of such insufficiency and defects in progress, materials, foreman, sub-contractors, or workmen, to the party of the first part; and if within three days thereinafter such insufficiency and defects are not remedied in a manner satisfactory to —the party of the second part, through the agency of said Superintendents, or otherwise, may enter upon the work, and suspend or discharge said party of the first part, and all employed under him, and carry on and complete the work by "days' work," or otherwise, as may elect, providing and substituting proper and sufficient materials and workmen; and the expense thereof shall be chargeable to the said party of the first part, and be deducted from any sum which may be due to him on a final settlement; and the opinion of said Superintendents shall be final, and their certificate in writing conclusive evidence between the parties hereto, on all questions and issues arising on or out of this fifth article of this Agreement, subject to the final decision of the referees hereinafter named.

6th. That the said part of the first part shall be solely responsible for any injury or damage sustained by any and all person or persons, or property, during or subsequent to the progress and completion of the works hereby agreed upon, from or by any act or default of the said part of the first part; and shall be responsible over the party of the second part for all costs and damages which said party of the second part may legally incur by reason of such injury or damage; and that the said part of the first part shall give all usual, requisite, and suitable notices to all parties whose estates or premises, being adjoining those upon which the works hereby agreed upon are to be done, may or shall be any way interested in or affected by the performance of said works.

7th. That the said part of the first part shall, from time to time, during the progress of the said works, apply to the said Architects for all needful explanations of the true intent and meaning of the said plans and specifications; and that "working-plans" shall, at the expense of the said part of the second part, be from time to time, and whenever requisite, furnished by the said Architects to the said part of the first part, upon reasonable notice being given to the said Architects that the same are requisite and needful; and further, that the said part of the first part will not and shall not, in the execution, performance, and fulfilment of this agreement, in any way deviate from the entire and exact compliance with, adherence to, and fulfilment of the said plans, "working-plans," and specifications, by reason of any practical difficulty which, in opinion, may or shall arise or occur; unless some such deviation shall, in the opinion and by the certificate of the said Architects, become absolutely necessary and unavoidable, in which case said part of the first part to make such deviation as they may be directed by said Architects.

And Whereas it is the intention of the parties hereto, that the said part of the first part shall bear and pay all the expenses necessary for and incident to the carrying into full and entire execution and completion all the works contemplated in this agreement, it is further understood and agreed by and between the parties to this agreement, that in case any lien or liens for labor or materials shall exist upon the property or estate of the said part of the second part, at the time or times when by the foregoing terms or provisions of this agreement a payment is to be made by the said part   of the second part to the said part   of the first part, such payment, or such part thereof as shall be equal to not less than double the amount for which such lien or liens shall or can exist, shall not be payable at the said stipulated time or times, notwithstanding anything to the contrary in this agreement contained; and that the said part   of the second part may and shall be well assured that no such liens do or can attach or exist before shall be liable to make either of the said payments.

It is expressly understood by the part   of the first part, that all the works described or referred to in the annexed specifications are to be executed by the said part   of the first part, whether or not the said works are illustrated by the aforesaid plans and working-drawings; and that said part   of the first part   to execute all works shown by the aforesaid plans and working-drawings, whether or not said works are described or referred to in the said specifications.

If any apparent discrepancy shall be found to exist between the plans, working-drawings, and the specifications, the decision as to the fair construction of said discrepancy, and of the true intent and meaning of the plans, working-drawings, and specifications, shall be made by the Architects hereinafter named; and said part   of the first part shall provide and execute the said works in accordance with said decision,—with the right of a final decision by the referees hereinafter named,—as a part of the original works undertaken by said part   of the first part

And Further Know all Men, That the parties hereto of the first part and of the second part severally, respectively, and mutually, hereby agree to submit, and hereby do submit, each, all, and every demand between them hereinafter arising, if any, concerning the value of any changes of, or omissions in, or additions to, the aforementioned plans or specifications, or concerning the manner of performing or completing the work, or the time or amount of any payment to be made under this agreement, or the quantity or quality of the labor or materials, or both, to be done, furnished, or provided under this agreement, or any other cause or matter touching the work, the materials, or the damages contemplated, set forth, or referred to, in or by this agreement, or concerning the construction of this agreement, to the determination of

the award of whom, or the award of a majority of whom being made and reported within   year from the time hereinbefore fixed upon for the final completion of this agreement to the Superior Court for the County of   the judgment thereof shall be final; and if either of the parties shall neglect to appear before the Arbitrator , after due notice given of the time and place appointed for hearing the parties, the Arbitrator may proceed in absence.

In Witness Whereof, The parties aforesaid have interchangeably set their hands and seals, the day and year first above written, to this and   other instrument of like tenor and date.
(*Signatures.*) (*Seals.*)

Executed and Delivered in presence of

STATE OR COMMONWEALTH OF
COUNTY OF A.D., 18 .

Then the above-named personally appeared and acknowledged the above instrument, by them signed, to be their free act.

Before me,
Justice of the Peace.

Specification to be annexed to the Building Contract.

Specifications of Materials to be provided and labor to be performed in the erection, and completion ready for occupancy (excepting plumbing and other water-works, painting, glazing, and piling) a block of houses for to be located on an estate recently purchased by him of on the easterly side of Street in within about 116 feet of the north-east corner of Street and Street. Said houses are to be constructed agreeably to plans prepared by , *Architect*, and under the direction of . , acting for and on behalf of said as superintendent of said building.

Description.—The block is to occupy and cover the full width from north to south of said estate, with its north and south ends located on the true boundaries of the estate (measuring about 117 feet in length, and just 45¾ feet in width). Said block is to be of four finished stories in height, besides a finished story within the intended French roof proposed to cover the whole structure. A cellar is to be constructed beneath the entire area of the building, and an area in the rear of the same; the latter to be of the form and dimensions indicated upon the drawings referred to. The clear heights of all the stories when finished are to be as follows, to wit: first, second, and third stories each 10 feet; and the French-roof story 9 feet. The cellar is to be 8 feet high in clear of the plastered ceiling and concrete flooring. The top of the flooring of the first story is to be located 3 feet 4 inches above the intended grade of the court-yard designed to be located in front or to the west of the block, as indicated particularly upon the profile drawing of the estate from east to west, forming one of the drawings before referred to; it being fully understood that the contractor for said block is to fill in, grade, and enclose with bank stone-walls, the north and south ends of the front or west yard of said estate, and the north, south, and east (or rear) yard walls of the said block, which walls are to be of the sectional form indicated by drawing of the same, forming one of the sets of drawings referred to.

Memo.—The front or west yard of the block will reach in width to the rear or east wall of a second block of tenement-houses designed to be erected by said Parker upon the front or westerly portion of said estate, but forming no part of the works to be estimated for under the specifications or plans.

Works.—The contractor for the block is at his own proper cost and expense to perform all labor of every kind requisite for its full completion, including all labor necessary for exterior grading, bank-walling, sewerage, flagging and paving, enclosing walls and fences, and for all other matters by these specifications required, and by the plans shown. Said works are to be of the best quality, and are to be performed by first-class workmen only, with the full right reserved to the said superintendent to discharge from the employ of the contractor for said block any workmen not of satisfactory capacity to him. Said works are further to be performed in such manner as to warrant and insure on the part of the contractor the most reliable and thorough construction, warranted in all cases to stand without start or flaw, and, in the case of all wood-work, warranted free from shrinkage, and so to remain. Said works are further to be so done as to progress at such rates of progress as are hereinafter stipulated, not, however, inconsistent with the quality of work required as aforesaid.

Materials.—All materials of every kind requisite for the full and entire completion of the block, together with its exterior adjuncts hereinbefore and hereinafter named, are to be provided at the sole cost of the contractors. Said materials are to be of the several kinds and quality hereinafter recited and described, but when not fully set forth in these plans and specifications, then the kinds to be used are in all cases to be the very best marketable qualities. All materials proposed to be used by them (the contractors) are at all times to be subject to inspection for approval or rejection by said superintendent; and the said superintendent shall be duly notified, and have the full opportunity in case he so elects to examine and inspect all materials before any of the same are delivered at the site of the building; and all materials he shall elect to reject shall be promptly replaced by such other stock as shall be satisfactory to the said superintendent, with the right on the part of the contractor to appeal from the decision of said superintendent to the referees named over the signatures of the owner of the property and the contractor for the block, in the agreement to be by them executed as a part of these presents. All materials designed for the building shall at all times be suitably housed, covered, and protected, including all walls daily on leaving the works. No window-frame or other exterior wood-finish shall be left unprimed more than one day after the same is worked or set. Any work or material damaged in any way during the erection of the building shall be promptly replaced on demand of the superintendent. The premises are not to be considered from Gloucester Place for the passage of men or materials, unless the written consent of the owners of the fee of said place is first obtained. The care and protection of the street (Washington) by day and night is not to be charged upon the contractors for said block; and, for this reason, all the materials of every kind designed to be used therein, must be landed fairly in the rear or to the west of the contemplated second or front block, with the right of passage, however, through the centre opening in said second or front block, for materials, and men engaged in the construction of said rear block.

Basement and Yard Drainage.—(See detailed plans of drains, cesspools, and aqueducts.) Three main drains of 16 inches clear diameter are to start from the three rear-yard cesspools, at proper levels of being wholly below basement-story flooring. These drains are to pass directly into and under the front yard of the block, after passing and connecting with three cesspools to be located on the basement-story centre passageway under the same, and in the said yard. They are to enter a single drain of two feet clear diameter; which drain the contractor for this block is to build through and under the archway of the contemplated front block of buildings, at proper levels, and with sure pitch, to connect with the Washington-street sewer in front of said Parker estate; which said connection is to be fully and legally made with said city sewer. But the cost of right to enter, including right to run plumbing works therein, will be arranged for and paid by said Parker. In addition to the three drains through the block aforesaid, there are to be branch-drains from the soil pipes of all water-closets, of 12 inches clear diameter each; and these drains are all to enter the principal drains aforesaid to the east or outside of the three cellar cesspools before referred to; and all other waste-pipes or sinks are to enter said drains to the east or inside of these basement-story cesspools. Eight aqueducts are to be laid from the shoes of the eight-roof conductors, and five others from the bottom of the five stone staircases outside of the basement. Three aqueducts may be square, but are to be fully six inches clear each way and are to be covered with 1¼ inch slate stones (not brick); and said aqueducts are to have full fall, workmanlike and endurable connections, with the other drains, all of which connections shall be in such localities as to make sure that no "soil" odor can "blow up" through the aqueducts into conductors or into areas at the foot of the several basement steps aforesaid.

Memo.—The paving of the yards and that of the center passageways inside of basement story is to pitch toward the several cesspools properly and regularly on inclines.

Memo.—Every wall and pier and wooden partition of basement story is to be lime-whitewashed (three heavy coats by an experienced expert). Proper aqueducts in brick are to be laid for Cochituate mains and metres, and for gas ditto ditto so far as the same may be required by superintendent to insure workmanlike construction for "entering" these matters from such points in the front yard of the block as the water and gas company bring same.

The two north and south boundaries of the front yard and three boundaries of the rear yard, excepting across the rear end of Gloucester Place, are to be fully enclosed with 12 inch brick walls resting on the copings of the several bank-walls, above which level (taken to be the front-yard level of the block), said walls are to be ten feet high. Said walls are to have in connection therewith buttresses of 8 by 16 inches each, from inside face of each wall; and the walls and the buttresses are to be capped with granite coping of 2 inches more width than the buttresses and walls, 4 inches thickness at the edges, and 9 inches in center, and to be straight and well tooled, and cramped on under sides, each piece to the other—all of which cramps are to pass down into the walls and buttresses. Said coping is to be wholly set in cement, and the whole of the joints flushed with same material. All yard paving is to be wholly in cement, and grouted and bedded in same manner as cellar paving aforesaid.

First Story. Brickwork.—The four exterior walls of this story are each to be 12 inches thick, and the two main, cross, party, subdivision-walls are to be of corresponding thickness with the outside walls. The two main corridor walls and those around stairways (three stairways) in this story are to be each 8 inches thick the entire length of the building, reaching fully in all cases to the top of flooring-planks of the second story. The twelve stacks of chimneys indicated on plans of this story are to be built in connection with and made part of the several walls, as shown. Said chimneys are to be commenced as floor-levels of the basement story upon stone-platform foundations to be made part of the other wall foundations, and built throughout said story with two piers of 20 by 20 inches each, to be covered with a semicircular arch tied with an iron beam bar, and the whole leveled up solid to first floor, with a flue in each chimney of 8 by 12 inches clear, square, and true, and plastered honorably over every square inch of inside surface, thick and heavy. No hearths or open fire-places are intended in chimneys. Water-closet flues, and the single flue of each room in which a chimney exists, is to be fitted with a 7 inch cast-iron funnel-piece and stopper of heavy and durable make; but no ventilating-flue is to be provided separate from the single smoke-flue of each apartment. All the said brickwork of the first story is to be laid in lime-mortar of first quality, Eastern stock, using sharp sea-sand only for same. All chimneys to have 8 inch backs and 4 inch withes.

Second and Third Stories.—The exterior walls are all to be continued 12 inches thick, and the chimneys built up in connection therewith in the same manner as before described for first story, with an additional flue of the second and third stories. The two interior, cross, division-walls will be carried through both these stories, but need be only eight inches thick The two enclosing walls of each of the two end stair-flights in both these stories are to be continued of brick, and of 8 inch thick each. The several window and door openings in all the walls of the three stories above the basement story are to be formed with reliable, arched heads on wooden lintels, and the exterior wall-windows to have full and square returns for window-frames. All frames are to be fitted in solid, and plastered in connection with brickwork. None of the walls are to be recessed beneath the windows. Every floor-plank is to be accurately levelled up, and the brickwork filled solid around it, and the roof-planks also at bottom. The fourth or French-roof story will have the four exterior walls built to top of plates of frame of roof, say 2½ feet above its flooring; and besides this the brickwork of the said four walls is to be continued up entirely to the roof-boarding under the gutter-flashing. The several corner quoins of the front side of the four corner pilasters of the side and the dentil course over the third-story windows of this side are all to be formed of brick; and all of them are to be made outside of the faces of the wall, thereby increasing in thickness as much more than 12 inches as the several matters project.

All chimneys are to be topped out, of one uniform height and one pattern; and this pattern is to be precisely like the detailed drawing to be given.

Memo.—The enclosing walls of the two end staircases are to be carried to roof-boarding of 8 inches thickness each.

Memo.—The nine cesspools hereinbefore referred to are to be 36 inches square in clear of walls; which walls shall be 8 inches thick, with an 8 inch bottom to same, and a four inch cut-off wall on iron bars, across the same. The whole inside to be rendered in hydraulic cement; and the curb and iron-trap strainer aforesaid to be set complete. The whole of the drains and aqueducts are to be most thoroughly rendered in hydraulic cement. The aqueducts may have 4 inch walls; but all the remaining drains shall have 8 inch walls, and shall be Gothic shaped at bottom; and the stone covering of said drains shall not be less than 2 inches thick, with full and square joints: the whole set in hydraulic cement. The walls of the drains shall be laid wholly in hydraulic cement. The contractor shall use all reasonable care that the grounds on which the drains, aqueducts, and cesspools to be built, is properly prepared to prevent settlement or start of said works; and, if the superintendent elects on account of the instability of the soil to substitute drain-pipe or plank drains for the brick ones hereinbefore stipulated, the contractor is to make the changes as directed; and all such difference of cost (more or less) as the superintendent elects to be just, shall be accepted by said contractor, and settlement made accordingly. Turn arches over all openings between cellar-piers, and level up to floors. The bricks to be supplied by the contractor are to be as follows, in quality: those for backing exterior walls, and for all interior walls and chimneys, may be of the Boston Brick Co.'s most costly cull; those for the drains and paving and other underground shall be Pilastow's Eastern or Charlestown clay brick, hand-made; the outside courses of the two end-walls and of the rear wall and of the chimney-tops shall be of same hand-made, even-colored, darkened, hard brick of uniform size, straight and true, and jointed-laid; the outside courses of the front wall shall be of a quality of face-brick as good and as fair a quality of Danvers face-brick, to be laid plumb-bond, and properly jointed off. All bricks shall be wet immediately previous to laying same. The contractor assumes all cost of supplying himself with Cochituate for use. The exterior cornices, brackets beneath, and small bank moldings beneath brackets, are all to be of wood, to be constructed and put up by carpenter; but the mason is to build in all brackets, and assist carpenter to space off and lay out same.

Slating.—The two upright sides of the roof are to be covered with 16 inch slates, Welsh; the whole to be of first quality, and agreeable to a sample which the superintendent will select, and submit to bidders before estimating. Said slates are to be put on with 2¼ inch lap (full), and to be truly bonded, to break joints in centers, to be put on with the heaviest quality of composition (not galvanized) nails. The chimney-tops; sides, tops, and sills of luthern windows; angle-corners of roof; top of upper wood-finish of roof; skylights; scuttle; scuttle over center staircase, or near it; as also all other required places,—are to be flashed with 10 oz. zinc and 4 lb. lead where the superintendent calls for the same; and the contractor for the slating is to be held responsible that furnishes and applies flashing-stock amply sufficient to insure an extra, first-class, tight, and permanent job, with every piece of stock cut and fitted and secured of such sizes and shapes as the superintendent, if he elects so to do, may direct.

Memo.—The slates of the front side of roof to have semicircular ends.

Gutters and Conductors.—The front and rear walls of the block, including the four heads or returns on the two ends of the block, are to be fitted with 20 oz. best quality sheet-copper to be of cima recta pattern, and made exactly in accordance with a full-size drawing to be given. This gutter is to be seated on to wood coving or casing of main cornice; and there is to be a back flashing from the inner edge of said gutter, on its top, of same quality and 16 oz. weight of copper, passing up beneath slating 8 inches, and passing under sills of each luthern window, and up to inside face to its top, and there turned on and secured with all suitable bends and heads of copper on each side of the lutherns, as well as over their entire top-surface or roof. The skylight-hatches, and that of the scuttle in flat of main roof, must be covered with 16 oz. copper also, and the whole made everywhere tight and secure and workmanlike. There are to be eight conductors of cold and rolled copper, of 16 oz. to the foot, put up, and firmly secured to the outside faces of the four exterior walls. Said conductors are to be connected with the gutters above by massive goose-necks most substantially soldered and secured, and of proper diameter; and the fifteen feet of said conductor, together with the shoes and underground lengths necessary for reaching and fully entering the aqueduct of brick, are to be made of the heaviest pattern of cast-iron, to be strongly connected with the four exterior walls, as to resist the most possible abuse that *boys* can bring to bear on said pipes.

Plastering. The walls, ceilings, and partitions of each of the four finished stories of the building, throughout every apartment, passageway, stairway, corridor, and hall, and including all closets and water-closets, are to be lathed on wood furring for five nailings, with sound, dry, pine-laths, free from sap and other defects, and secured with 3 penny nails. The laths to be universally a full quarter of an inch apart. The ceilings of the cellar to be lathed for plastering throughout. Each floor of the four finished stories is to be plastered between upper and under with a heavy coat, ¼ inch thick, of lime and hair mortar. All other plastering is to be done two coats,—one of lime and hair mortar, and the second a skim coat of lime and sand putty; forming the first quality of two-coat work, as usually understood in best houses, as the walls are not to be papered. The ceilings and walls both are to be finished of entire uniform shade of plastering, without staging-streaks, or break-offs in any place. No corner or centre-pieces are required. The contractor shall do the usual and fair amount of patching after carpenters have finished, without charge to owner of the building. The risk of the plastering being touched by frost, if work of building is delayed, rests with the plasterer wholly.

Miscellaneous.—Mason. In both parlor and kitchen of each tenement, there is to be a red slate-stone mantel, to be supported by two iron bronzed brackets of some neat pattern, the whole to be selected and approved by the superintendent. The mason is to include the paving of the whole area of the yard in front of the block up to the rear line of the contemplated front block of houses; and said paving is to be done in cement, like that hereinbefore required.

Carpentry.—The carpenter is to be equally responsible with the mason that all parts of the building are correctly laid out, from the several plans by the architect; and he is, in consultation with the superintendent and mason, to arrange all details and portions of construction in ample season for them all to be applied correctly to the buildings. He is also at his own cost to prepare all centres, not only for windows and openings, but also for drains. He is also to make all necessary poles and rods as guides for laying out all works. He is to make skeleton frames, and set the same, for all openings in walls. He is to cover all freestone and granite projections, including doorways, and water-table of underpinning. He is to safely shore all floors, under all such points as the superintendent directs, while the skeleton of the structure is in progress. He is to make one set of patterns from the full-size drawings of all freestone, molded, and arch work. His works are to embrace all branches of trades hereinbefore stipulated under the head of work and labor and materials, it being understood that in connection with the contractor for the masonry, the buildings are to be left in a completed state, ready for occupancy, excepting only metal-works of the plumbing. No furnaces, fireplaces, grates, stoves, or heating-apparatus of any kind, being intended to be required of the contractors, saving only chimneys, funnel pieces, and stoppers.

No papering is to be required of contractors; and no gas piping or fixtures is to be embraced in the estimates of contractors. Such of the water-closet ventilators as are required of wood are to be constructed and topped out, and otherwise fully put up and completed, precisely as superintendent says.

Framing.—To provide the first marketable quality of Eastern spruce stock, and frame, put on, and otherwise fully complete, the floors of the first, second, third, and fourth stories, with planks of 2 by 12 inches, to be placed as indicated by flooring-plans; spanning in all cases from the front and rear exterior walls on to the corridor-walls, which run through the centre of the length of the entire building. Each floor is to contain headers and trimmers of 4 by 12 inches wherever indicated by the plans, excepting those for enclosing staircases, which are, in all the floors, to be 6 by 12 inches. The planks in all the floors over the centre corridor may be 2 by 9 inches only. The first floor will contain girders of 7 by 10 inches, to be located in the position indicated by the flooring-plan of that story. These girders are to be of the soundest white pine, of last year's growth, and last year's delivery in Boston, and not in water for the last six months at least. These girders are to be worked square and true, and are to rest on the exterior walls and interior piers. Each flooring is to have four full rows of diagonal bridging of inch-board pieces 3 inches in width and 1 inch thick, to be accurately cut in, and nailed with twelve pennys. The whole of the flooring-planks are to rest just one full half-brick in length of bearing on walls, and four inches full on the corridor walls and partitions; and the same of the headers and trimmers in each floor. All headers and trimmers are to be mortised and tenoned and oak-pinned, and those of the stairways are to have wrought-iron stirrup-straps of 2 by ⅜ inch iron. The upper and under edges of every flooring-plank is to be worked by a plane to a regular crown of ¼ of an inch in their length. There shall be twelve wrought-iron ties attached to the trimmers of each floor in the position the superintendent shall say; and all these ties are to go to, and be "upset" in, the exterior walls to within 4 inches of the outer face of each wall. Each tie to be 3½ feet long, of ⅞ inch round iron, in addition to the length required for "upsetting" the two ends.

The roof to be framed with its two upright, angular sides of plank 3 by 9 inches, to be placed only 18 inches apart on centres. Said planks are to be footed, and securely spiked to wall-plates of 3 by 10 inches; which plates are to be bedded on and bolted to the exterior walls by bolts being built in for the height of 5 feet in said walls once in every 15 feet length thereof. The tops of the aforesaid rafters are to be headed into a border-stick, which is to extend the entire length of the two sides of the block, and is to measure 5 by 9 inches; being properly framed (not merely spiked) on to the rafters. The border-piece and the heads of the two main corridor-partitions are to form supports for the two ends of the planks designed to form the top or flat portion of the roof. Said planks are to be fully 3 by 12 inches, to be placed only 8 inches apart on centres, and bridged precisely like the floors aforesaid by with one row only on each side of the corridor-partitions. The roof-stock is all to be as dry and as perfect as that for the floors aforesaid; and the upper edges on outer edges of all the planks are to be worked true with plane, and those in the flat to be crowned regular 1 inch in their length. Every part of the framing of floors and roof is to be so mortised, tenoned, spiked, nailed, stayed, and otherwise finished and secured, so as to make not only a first-class, workman-like job, but one to be warranted free from start or tremble, and permanently so to remain. On each side of each luthern window, there is to be a stud of 3 by 6 inches, with a head-piece of same size at top of window; and these six studs are designed to go perpendicularly down to the top of the roof-story flooring, just down the exterior walls, and there to foot on a plank which is to run the whole length of the building; which plank, as well as the side-studs and head-piece, are all to be firmly spiked and secured.

Furring and Partitions.—The brick walls, ceilings, and stairway throughout the four finished stories, are to be furred with 3 by 1 inch dry spruce furrings, set to give five nailings to a lath. They are to be put on the walls with twelvepenny nails, and on the ceilings with tenpennies. Grounds ⅞ of an inch thick are to be put up for all finish, and ⅞ inch beads for the angles of the walls and stairways.

The partitions, except those which are brick, are to be framed with sound, seasoned spruce lumber; the studs to be 2 by 4 inches; door studs and girths, and window studs and girths, 3 by 4 inches; plates 3 by 4; and sills

2 by 4 inches: all to be thoroughly bridged with cross bridging, and to be braced over the doors and windows.

All of the above work is to be done in the most thorough manner, and, when ready for the plastering, is to be plumb, square, and straight.

Memo.—The caps and sills of every partition in every story are to be seasoned Southern pine, properly fitted and secured.

Tinning.—The dormer-window roofs, and the upper portion or flat of the main roof, are to be covered with best quality of charcoal-leaded, of first quality MF brand roofing-tin; to be laid, lapped, soldered, and secured in the most thorough manner, and warranted a first-class and permanently tight job throughout.

Rough Boarding.—The roofs are to be boarded, and the under-floors to be laid with sound, seasoned white-pine boards, matched and mill-planed; laid close, and thoroughly nailed; and those to the slated portion of the roof are to be covered with the best quality of tarred sheathing-paper.

Outside Finish.—The dormer-windows, cornices, brackets, and small band-moldings beneath them, are to be wrought of thoroughly-seasoned, clear, white-pine stock, in the forms shown by the drawings; and they are to be thoroughly secured to the brickwork where they come in contact with it.

The doorway is to be framed with 2 by 4 inch studs, and 2 by 6 inch rafters, and is to be boarded with matched and mill-planed pine covering-boards, and covered with tin, like the roof. It is to have a rebated plank-door-jamb, 4 inch outside and inside casings, and a white-pine door with four plain panels. The door is to be 2 inches thick, hung with stout, loose butt-hinges, and fitted with a good lock, inside bolts, and neat and durable trimmings.

Windows.—All the windows inside and out, excepting those in the cellar, are to have box-frames with 2 inch sills and yokes, and 1 inch inside, outside, and back casings; and staff-beads of white pine for those in the brick walls; but no back casings or staff-beads for those in the wooden partitions. They are to have 1 inch pulley-stiles, $\frac{1}{2}$ inch inside, and $\frac{1}{4}$ inch parting beads of hard pine.

Each of the above windows is to be fitted with two $1\frac{1}{4}$ inch white-pine sashes, molded and coped. The lower sashes in the inside of partition-windows are to be firmly secured to the frames; the upper sashes in the said windows, and both sashes in each of the other windows, are to be hung with best flax sash-lines, steel axle-pulleys, and round iron counter-weights, and fitted with bronze sash-fastenings to cost $7 per dozen. They are to have pockets neatly cut into the pulley-stiles, and secured by brass screws. Each window is to be cased as shown by the drawings, and finished with molded stools and molded architraves, as therein represented. The upper sash of each and every window in all the halls and staircases is invariably to be hung and fastened.

The cellar-windows are to have white-pine rebated plank-frames, and a single sash each. The sashes to be hung with stout iron hinges, and fitted with neat and durable buttons and catches.

The skylight frames are to be of thoroughly-seasoned, clear, white-pine stock, rebated for the sashes, put together with white lead, and finished off in a neat and durable manner.

Doors.—All the doors are to be made of thoroughly-seasoned, clear, white-pine stock; the outside doors to both front and rear being 2 inches thick, the principal doors in the rooms and entries 1¾, and the closet doors 1¼ inches thick. The outside doors are to be made in the forms shown on the drawings; are to be hung with three sets of 5 inch, ornamental bronzed, loose butt-hinges, and fitted with locks, bolts, and trimmings, to be selected by the superintendent, and to cost for such locks, bolts, and trimmings, the sum of $6 exclusive of the cost of putting on. The basement doors are to have locks, trimmings, bolts, and loose butt-hinges, to cost $5 to each door. The doors to the entries, rooms, and closets, are to have four molded panels to each, and are to be of the sizes marked on the plans. All are to be hung with stout, iron, loose butt hinges. Those for the storerooms, pantries between the different rooms, and the entry doors, are to have locks and trimmings to cost $5 to each door, on the average. The doors to the bedrooms, closets, and to the water-closets, are to have mortised spring-latches with knobs, etc., to correspond to those to the other doors; and each water-closet is to be fitted with an inside brass bolt, neat and durable. The doors to the coal-bins are to be made of matched and mill-planed white-pine stock, battened; are to be hung with stout strap-hinges; and each is to be fitted with a

The fly-doors of the vestibule are to be 2½ inches thick, with plain panels. They are to be hung with loose butts, double action springs of a satisfactory quality, brass bolts to the top and bottom of one half, and a lock to the other half. This door, or the outside door, at the option of the superintendent, is to have a lever night-lock of good quality, with fifty (50) keys.

The inside doors are to be finished with hard-pine thresholds, 2 inch rebated and beaded frames of white pine, and architraves to correspond with the window-finish in the various parts of the building.

The outside doors are to be hung to 3 inch plank frames, properly dogged to the thresholds; and jambs finished inside like the inside door, and outside with staff-molding.

Blinds.—Each window (excepting those in the basement and French roof) on the exterior of the building is to have a pair of 1¼ inch mortised slat-blinds, made with rebated and beaded stiles, and three rails to each. They are to be hung with the best quality of blind-hinges, and fitted with satisfactory fastenings.

Stairs.—The stairs are to be framed with deep spruce-plank stringers and landings and winders, as shown on the drawings. They are to have white-pine string and gallery finish, hard-pine risers, treads, and balusters. The balusters to be round, and 1¼ inches in diameter. The posts are to be 10 inches square, and the newels 5 inches. They are to be molded and capped, and the post panelled as per drawings. The rail is to be 3½ inches in width, and of a satisfactory pattern. The posts, rails, and newels are to be of thoroughly-seasoned black walnut; and the rails are to be not less than 3 feet high. The stairs to the cellar are to be framed with plank stringers, and to be finished with planed pine-plank risers, and hard-pine treads, and plank hand-rails and supporters.

Dado and Inside Finish.—The walls of the entries throughout the four finished stories, and of the kitchens and water-closets throughout the building, are to be dadoed to the height of 3½ feet above the floor with narrow matched and beaded white-pine sheathing finished with a molded capping of the form of the stool nosing.

The walls of the parlors and bedrooms are to have molded bases 10 inches high, and 1¼ inches thick. The other walls are to have levelled bases 8 inches high, and ⅞ of an inch thick.

The water-closets are to be finished off with black-walnut stock, the covers and seats being hung to raise, and all woodwork being put up with brass screws. Ventilating boxes or flues of brick are to be made for the water-closets where indicated by the drawings, carried out through the roof, and finished in a neat and durable manner.

All the inside woodwork not otherwise specified is to be wrought of thoroughly-seasoned, clear, white-pine stock, free from shakes and sap, and put in in the best and most workmanlike manner.

Closets.—Each pantry and china-closet is to be fitted with a case of four drawers made in a neat and substantial manner. One set of drawers in each tenement to have strong tumbler-locks, and each drawer to have two drawer-pulls.

These closets are to have shelves and cupboards as directed, and each is to have cleats of cast-iron (single) hooks.

The bed-room closets are to have cleats of double cast-iron clothes-hooks placed 6 inches apart on three walls of each, and are to be shelved round over the clothes-hooks. The cupboards above mentioned are to have brass thumb-slides: strong tumbler-locks and drawer-pulls.

Floors.—The floors to the halls and corridors are to be laid with thoroughly seasoned, clear, hard-pine stock, not exceeding 5 inches in width, laid close, and thoroughly nailed and smoothed. All the other floors in the four finished stories are to be laid with thoroughly-seasoned, kiln-dried, spruce floorings, selected for clearness and soundness. They are not to exceed 6 inches in width, and are to be laid close, thoroughly nailed and smoothed, and put down as soon as taken from the dry-house.

Sinks.—Each kitchen is to have a soapstone set in a pine-plank frame. The sinks are to be 3 feet long, and 1 foot high, and 18 inches wide inside, and are to be finished beneath in a neat and durable manner, with cupboards. They are to be backed up with pine, and fitted to receive the plumbing. Each sink is to have a composition cesspool.

Coal-Bins.—There are to be coal-bins finished off in the cellars, one for each tenement. Each bin is to be fitted inside the door with two separate compartments capable of holding 1 ton of coal to each compartment, and with another to take 2 barrels of kindlings. The exterior woodwork is to be of pine, mill-planed, and the interior partitions of spruce; these latter being fitted with sliding gates, and boxings around them to keep the coal from the floor. All the above work is to be done in the most thorough and workmanlike manner.

Bells.—The outside door to each tenement is to be fitted with a bell leading to the kitchen. It is to have a handle to correspond with the doorknobs. Each tenement is to have a bell to the porter's room, fitted with a bronze slide. All the above are to be gong-bells with tubed wires, and put in the most perfect manner.

GLOSSARY

OF TERMS USED IN ARCHITECTURE.

ABACUS. The upper plate upon the capital of a column, upon which the architrave rests. There are several varieties, as the Grecian, Roman, and Corinthian Doric.

ACANTHUS. An ornament resembling foliage; used in the capital of the Corinthian and Composite orders.

AMPHIPROSTYLE. An edifice with columns in front and behind, but not on the sides.

AMPHITHEATER. A circular edifice, having rows of seats, one above another, inclosing an open space called an arena; used for combats of gladiators, and other public sports.

ANNULET. A small flat fillet encircling a column. It is several times repeated under the Doric capital.

APOPHYGES. A hollow molding. That part of a column where it springs out of its base.

ARCADE. Arches supported by columns, either open, or backed by masonry. A long arched building, or a single arched aperture, sometimes called a vault.

ARCH. A construction of bricks, wood, or stone, disposed in the form of a curve. There are several parts, as the keystone, which enters the top of the arch like a wedge, binding the work. Springers, the bottom stones which rest on the supports; and span, which is the distance across the arch.

ARCHITECTURE. The art or science of building; especially the art of constructing houses, bridges, and other buildings, for the purposes of civil life.

ARCHITRAVE. 1. The lower division of an entablature, or that part which rests immediately on the column. 2. Also, the ornamental molding around the exterior of an arch. 3. A molding above a door or window, and the like. 4. This term is also applied to door and window casings.

Arris. The edges formed by two surfaces meeting together, whether plain or curved. In stucco work, when two surfaces meet, as the corner of a beam or cornice, this term applies.

Arris Fillet. A triangular piece of wood laid against a chimney or wall, to raise shingles or slates, to throw off the rain.

Astragal. 1. A little round molding, which surrounds the top or bottom of a column. 2. Also, often used in the capital of the Ionic column. And it is also used for various purposes in common work.

Baluster. A small column used as the support to the rail of a stair-case.

Balustrade. A row of balusters topped by a rail, inclosing balconies, altars, etc.

Balcony. A platform or projection from the outside walls of a house, generally inclosed by a balustrade.

Baldachin (Bal-da-kin). A structure resembling a canopy, sometimes supported by a column, generally placed over an altar; also, over doors and windows.

Band. Any flat low molding, broad, but not deep.

Bartizan. The small overhanging turret which projects from the angles of towers, and other parts of a building.

Base. 1. The lower projecting part of a room, consisting of the plinth and its moldings. 2. The part of a column between the top of the pedestal and bottom of the shaft.

Baston. A round molding in the bottom of a column; called, also, torus.

Battlement. A notched parapet, originally used only in fortifications, but afterwards in buildings.

Bay-window. A window projecting outward from the walls.

Bond. The union of bricks or stone in a wall, so as to bind them together. This is done in several ways; as the English bond, where one course consists of bricks, with their ends toward the face of the wall, called headers; and the next course, with their lengths parallel with the face of the wall, called stretchers. Flemish bond, where each course consists of headers and stretchers, alternately, so as always to break joints. These bonds are now nearly dispensed with, and have given place to the running bond, in which all bricks are stretchers, and are bound together by strips of sheet-iron running crosswise. The plumb bond is that in which the joints of alternate courses are directly over each other.

Bracket. A piece of wood, stone, or metal, projecting from a wall, to support shelves, statuary, or other objects.

Buttress. The projecting support to the exterior of a wall, most commonly applied to churches in the Gothic style; but also to other buildings. It also implies a support.

CANOPY. An ornamental projection, in the Gothic style, over doors and windows.

CANTILEVER. A projecting block, or bracket, supporting a balcony; the upper member of a cornice, the eaves of a house, etc.

CAPITAL. The head or upper part of a column or pilaster. There are six varieties, each adapted to its respective order, namely: The Gothic,—ornamented with leaves and foliations. Composite,—the Ionic grafted on the Corinthian, called also the Roman or Italic capital. Tuscan,—plain and unornamented, and much like the Doric. Ionic,—whose distinguishing feature is the volute of its capital, and is less ornamented than the Corinthian. Doric,—much like the Tuscan, and between that and the Ionic in ornamentation. And the Corinthian,—distinguished for its profusion of ornaments.

CARPENTRY. The art of framing and joining timbers in the construction of buildings.

CATHERINE WHEEL WINDOW. An ornamental circular window, with radiating divisions, or spokes.

CAVETTO. A hollowed molding, whose profile is the quarter of a circle.

CHANCEL. That part of a church between the altar and balustrade that incloses it.

CHAPTREL. The capital of a pier or pilaster which receives the arch; called, also, an impost.

CEILING. The upper interior surface, opposite the floor.

CHAMFER. To cut in a sloping manner, or to bevel a square edge.

CHAPLET. A little carved molding.

CINQUE-FOIL. An ornamental foliation, having five points or cusps, pointing toward the center; used in windows and panels, and often frescoed on walls.

CHOIR. Place occupied by the officiating clergyman. The chancel.

CLUSTERED COLUMN. A column which is composed, or appears to be composed, of several columns collected together.

COLUMN. A cylindrical support for a roof or ceiling, and composed of base, shaft, and capital.

COMPOSITE ORDER. An order of architecture made up of the Ionic order grafted on the Corinthian; called, also, the Roman or Italic order.

COMPOSITE ARCH. A pointed, or lancet arch.

CONSOLES. A bracket or shoulder-piece, or an ornament on the key stone of an arch, and sometimes used to support cornices.

COPING. The highest or covering course of masonry in a wall.

CONCRETE. A mass of stone chippings or pebbles, cemented by mortar, and used for foundations, cellar floors, etc.

CORONA. A large flat member of a cornice, to carry off the rain that falls on it; called, by workmen, the drip.

CORINTHIAN ORDER. An order characterized by its profusion of ornaments, having eight angular volutes; also, eight smaller ones, called helices.

CORBEL. 1. The base or tambor of the Corinthian column, so called from its resemblance to a basket. 2. A short piece of timber or iron in a wall, jutting out as occasion requires, in the manner of a shoulder-piece; the under part is sometimes cut in the form of an ogee, a face, or other figure.

CORBEL TABLE. A projecting course of masonry, used in supporting a battlement or cornice, and resting on corbels.

COVE. An arch overhead, where ceilings connect with the walls.

CORNICE. A molded projection which crowns or finishes the part to which it is attached; as the cornice of an order, of pedestal, of a door, window, or house.

CROCKET. An ornament formed in imitation of carved and bent foliage, and placed upon the angles of spires, canopies, etc.

CURB ROOF. 1. A roof having a double slope, and composed on each side of two parts having unequal inclinations. 2. Also, a gambrel roof; a mansard roof. 3. In the curb roof, the rafters in both inclinations are straight.

CURB PLATE. Plate in a curb roof, that receives the feet of the upper rafters.

CUPOLA. A spherical or dome-like vault on the top of an edifice or building; usually on a tower or steeple or public building.

CYMA. A member or molding, the profile of which is wave-like in form

CYMA RECTA. A molding, hollow in the upper part, but swelling in the lower part.

CYMA REVERSA. A molding rounding in the upper part, but hollowing in the lower part.

DADO, or DIE. The part of the pedestal of a column between the base and cornice. Also, that part of an apartment between the base and surbase; or panel between the plinth and impost molding.

DENTIL. An ornamental square block or projection in cornices, bearing some resemblance to teeth; used particularly in the Ionic, Corinthian, and Composite orders.

DOOR-STOP. One of the pieces of wood against which a door shuts in its frame.

Door Case. The frame which incloses a door.

Doric Order. Belonging to the second order of columns,—between the Tuscan and Ionic. The Doric order is distinguished for strength and simplicity.

Dormer Window. A window placed on the roof of a house; the frame being placed vertical on the rafters.

Echinus (e-ki-nus). A molding of the same form as the ovolo or quarter round; but properly so called only when ornamented or carved with eggs and anchors.

Engaged Column. Column sunk partly into walls to which they are attached, and standing out, at least, one-half their thickness.

Entablature. That part of an order over the columns consisting of architrave, frieze, and cornice.

Façade. A front view or elevation of an edifice.

Fascia. A flat member of an order or building, like a flat band or broad fillet. The member attached to the ends of the rafters, directly under the eaves of buildings.

Festoon. An ornament of carved work, representing a wreath of flowers, fruits, and leaves intermixed or twisted together, and represented as hanging or depending in an arch.

Fillet. A little square member or ornament used in divers places, but generally as a corona over a greater molding; sometimes as a small square under other moldings. Also, the square part of the cyma recta and ogee moldings.

Finials. A knot or bunch of foliage that forms the upper extremities of pinnacles in Gothic architecture.

Flute. A channel in a column or pilaster; it is used chiefly in the Ionic and Doric orders; sometimes in the Corinthian and Composite, but rarely in the Tuscan.

Foliation. The act of enriching with feather ornaments, or the ornaments themselves.

Fret. An ornament consisting of small fillets, intersecting each other at right angles.

Fresco. A method of painting on walls, performed with mineral and earthy pigment, on freshly laid stucco ground of lime and gypsum.

Frieze. That part of an entablature between the cornice and architrave, often enriched with figures of animals and other ornaments; whence its name.

Furring. The act of nailing strips of narrow board, usually 1 × 3, to the floor timbers to make a level surface for laths.

Gable. The vertical triangular end of a house or other building, from the eaves to the top.

GABLE ROOF. The sloping roof that forms the gable.

GIRDER. The principal piece of timber in a floor, girding or binding the others together.

GOTHIC. A style of architecture with high and sharply-pointed arches, clustered columns, etc.

GROINED ARCH. An arch having an angular curve, made by the intersection of two half cylinders or arches, as a groined ceiling.

GUTTER. A channel at the eaves of a roof to carry away the rain.

HAMMER BEAM. A beam acting as a tie at the feet of a pair of the principal rafters, but not extending so as to connect with the opposite side of the building. It is generally supported by a rail, springing from a corbel below, and also itself supports another rail,—forming, with that springing from the other side, an arch. Often used in churches built in the Gothic style.

HANCE. The end of an elliptical arch, which is the arc of a smaller circle than the arch itself.

HANGING BUTTRESS. A buttress supported on a corbel, and not resting on the solid foundation.

HAUNCH OF AN ARCH. The parts between the crown and springing.

HELIX. The little volute under the flowers of the Corinthian capital.

HIP KNOB. An ornament placed upon the roof of a building, either upon the hips, or at the point of the gable.

HIP MOLDING. A molding on the rafter or beam which forms the hip of a building.

HIP ROOF. A roof having sloping ends and sloping sides.

HOOD MOLDING. A projecting molding over the head of an arch.

INTERLACING ARCHES. Arches usually circular, so constructed that their curves intersect.

IONIC ORDER. An order whose distinguishing feature is the volute of its capital, of which it has four. The column is more slender than the Doric and Tuscan, but less slender and less ornamental than the Corinthian and Composite.

JAMB. The side-piece or post of a door or window, or any other aperture in a building.

JOINERY. Art of one who covers or finishes buildings. The joiner commences where the carpenter finishes.

KERF. To saw or notch in wood, to make it flexible or easily bend.

KING-POST. A post rising from the tie-beam to the roof.

LATTICE. Any work of wood or iron made by crossing laths, rods, or bars, and forming a network.

LANCET WINDOW. A high and narrow window pointed like a lancet.

LEAVES. Ornaments of carved work resembling leaves of certain trees, forming part of ornamental capitals.

LINTEL. A longitudinal piece of wood or stone placed over a door, window, or other opening. A head-piece.

LISTEL. A little square molding. A longitudinal ridge between the flutings of the Grecian column, except the Doric.

LOTUS. An ornament used on Egyptian columns, resembling a lily.

LOUVER WINDOW. An opening in a bell-tower church steeple, crossed by a series of bars, or sloping boards, to exclude the rain, but allow the passage of sound from the bells.

LOZENGE MOLDING. A molding used in Norman architecture, characterized by lozenge-shaped compartments.

LACUNARS. Spaces sunk or hollowed into a ceiling.

MANTEL. The work over a fire-place in front of a chimney; especially a narrow shelf above the fire-place. Called, also, mantel-piece.

MASONRY. The occupation of a mason, as the building of stone and brick walls, or laying stone and bricks in mortar in the construction of buildings.

METOPE (met-o-pe). Spaces between the triglyphs of the Doric frieze.

MINARET. A slender lofty turret, or Mosques of Mohammedan countries.

MITRE. This term is applied to pieces meeting at any angle, and matching together so as to make a joint.

MODILLION. The enriched block, or bracket, found under the cornice of the Corinthian entablature, and sometimes less ornamental in the Ionic, Composite, and other orders.

MOLDING. A projection beyond the wall, column, wainscot, etc., an assemblage of which form a cornice, a door-case, or other decoration. The names by which they are commonly known are as follows: astragal, ogee cymateum, cavetto, scotia or casement, apophyges, ovolo or quarter round, torus, reeding, band, etc.

MORTISE JOINT. A joint made by a mortise and tenon.

MULLION. A slender bar which forms the divisions between the lights of windows; also, one of the divisions in paneling.

MUTULE. A projecting block worked under the corona of the Doric column, in the same situation as the modillion in the Corinthian and Composite orders.

NAVE. The middle or body of a church, extending from the choir, or chancel, to the principal entrance; also, the part between the wings or aisles.

NEWEL. The upright post about which the steps of a circular staircase winds.

NICHE. A cavity or recess, generally within the thickness of a wall, for a statue, bust, or other erect ornament.

NOSING. That part of the step-board of a stair that projects over the riser; also, any like projection.

OGEE. A molding, consisting of two members, one concave, the other convex; or, a round and a hollow.

OGEE ARCH. A pointed arch, each of the sides of which is formed by two contrasted curves.

ORIEL WINDOW. A large bay window in a hall or chapel, sometimes resting on a corbel.

OVOLO. A round molding; the quarter round.

PATEREE. A circular ornament resembling a dish, often worked on friezes.

PAVILION. A temporary movable habitation or tent. Sometimes a pavilion is a projecting part in the front of a building; sometimes it flanks a corner.

PEDESTAL. The base or foot of a column. It consists of three parts,—base, dado or die, and cornice.

PEDIMENT. The triangular facing of a portico, or a similar decoration over doors, windows, gates, etc. Also, in other buildings, where the eave cornice returns across the end to meet the opposite cornice.

PENDANT. A hanging ornament, used in roofs, ceilings, etc.; much used in Gothic architecture.

PENTASTYLE. An edifice with five columns in front.

PERISTYLE. A building encompassed with a row of columns in front.

PILASTER. A square column usually set within a wall, and projecting a fourth or fifth part of its diameter. The base, capital, and entablature of a pilaster are the same as those of a column.

PINNACLE. A slender turret, or part of a building elevated above the main building.

PITCH OF A ROOF. The inclination, or slope of the sides, usually expressed in parts of the span, as a third pitch or fourth pitch. By these expressions, the height of the top of the rafters above the plate is a third or a fourth of the width of the building. Sometimes by the length of the rafters in parts of the span, as two-thirds pitch, or three-quarters pitch, etc.; in this case, the length of the rafters is two-thirds or three-quarters of the width of the building. Also, the Gothic pitch,—that in which the rafters equal the span. Elizabethan pitch,—that in which the rafters are longer than the span. Grecian pitch,—in which the rafters are one-ninth to one-seventh of the span. And the Roman pitch,—in which the height is one-fifth to two-ninths of the span

PLATE. A piece of timber which supports the ends of the rafters.

PLINTH. A square, projecting, vertically faced member, forming the lowest division in the base of a column or pilaster; also, used in other positions.

PLANCEER, or PLANCHER. The under side of a cornice; a soffit.

PODIUM. Balcony, or open gallery.

PORCH. A kind of vestibule at the entrance of temples, churches, halls, and other buildings; hence, an ornamental entrance way.

PORTICO. In modern usage a covered space, inclosed by columns, at the entrance of a building.

PURLINE. A piece of timber, extending from end to end of a building or roof, across and under the rafters, to support them.

PUTLOG. A piece of timber on which the planks of a stage are laid, one end resting on the ledger of the stage, and the other in a hole in the wall, left temporarily for the purpose.

QUARTER-FOIL. A figure disposed in four segments of a circle, resembling an expanded flower.

QUEEN-POST. One of the suspended posts in a truss-roof, framed below into the tie-beam, and above into the principal rafter.

QUIRK. A small acute channel, by which the rounded part of a Grecian ovolo or ogee molding is separated from the fillet.

ROOFS. These may be divided as follows: Gable,—straight rafters. Hip,—all sides inclined. Conical,—shaped like a cone. Ogee,—a round and a hollow. Shed,—roof on one side. Curb,—two inclinations on each side. M—two buildings whose roofs form the letter M. French,—two sets of rafters on each side, of unequal slopes; the upper ones generally straight, the lower ones curved inwardly.

RABBET. A cut made in the edge of a board, so that it may form a joint with another board similarly cut, by lapping, and forming an even surface.

RAIL. Horizontal part in any piece of frame-work or paneling.

RAKE. Pitch or inclination of a roof.

RAMP. A concave bend in any piece of ascending or descending workmanship.

RECESS. Falling back of the walls in a room. Also, the space made where doors or windows are set back or forward from the line of the wall.

READING. A small convex molding.

RETURN. The continuation of a molding or projection in a different direction. Also, the side or part that falls away from a straight line.

SCARF JOINT. Joints formed by scarfing the ends of two pieces of timber, so that they will compare.

SEAT OF A HIP. A level line over which a hip rafter stands.

Scotia. A concave molding used in the base of a column, between he fillets of the tori, and in other situations; its outline is a segment of a circle, often greater than a semicircle.

Scroll. A convolved or spiral ornament. The volute of the Ionic and Corinthian capitals.

Soffit. Under side of stairways, archways, entablatures, cornices, or ceilings.

Spire. A body that shoots up in a conical form. A steeple.

Spandrel. An irregular, triangular space between the curve of an arch and the including right angle. Often seen in fresco paintings.

Stall. A small apartment, where merchandise is exposed for sale; as a butcher's stall.

Stile. Upright piece of framing or paneling.

Stool. The flat piece upon which a window shuts down, and which corresponds to the sill of a door.

Stucco. Plaster of any kind used as a coating for walls, especially a fine plaster composed of lime or gypsum with sand and pounded marble; used for internal decorations and fine work.

Straight Arch. A form of an arch in which the intrados are straight; but with its joints drawn concentrically, as in the common arch.

Surbase. Cornice, or series of moldings on the upper part of a pedestal; sometimes called a chair-rail.

Table. A smooth, simple ornament of various forms, most usual in that of a long square.

Taxis. The disposition which assigns to every part of a building its just dimensions.

Tie-beam. A beam acting as a tie, as at the bottom of a pair of the principal rafters, and prevents them from thrusting out the walls.

Torus. A large molding used in the base of columns. Its profile is semicircular.

Tower. A lofty building much higher than it is broad; either standing alone or forming a part of another edifice, as of a church, castle, and the like.

Threshold. The door-sill. The plank, stone, or piece of timber or board that lies at the bottom or under a door of a dwelling-house, church, or the like.

Transom. A horizontal cross-bar over a door or window, sometimes used for the purpose of supporting a sash over a door.

Tracery. An ornamental divergence in the mullions in the head of a window, into arches, curves, and flowing lines.

TRELLIS. A structure or frame of cross-barred work, for various purposes.

TRANSEPT. A part of a church at right angles to the body of the church. In a cruciform church, it is one of the arms of the cross.

TRIGLYPH. An ornament in the frieze of the Doric order, repeated at regular intervals. It consists of a piece of board or other material, rectangular in shape, with two parallel slots running from the bottom nearly to the top, nearly dividing the piece into three parts, called glyphs.

TURRET. A little tower or spire attached to a building, and rising above it.

TUSCAN ORDER. A capital of a Tuscan column is plain, unornamental, and much like that used in Doric architecture.

VERANDA. A kind of open portico, formed by extending a sloping roof beyond the main building.

VESTIBULE. The porch or entrance into a house. A hall or antechamber next to the entrance, and from which doors open to the various rooms in the house.

VOLUTE. A kind of spiral scroll used in the Corinthian and Composite capitals, of which it is the principal ornament. The number of volutes in the Ionic capital is four; in the Composite, eight. There are also eight angular volutes in the Corinthian capital, accompanied by eight smaller ones, called helices.

WELL-HOLE. The open space in the middle of a staircase, beyond the ends of the stairs.

DEFINITIONS

OF THE MOST COMMON TERMS IN DRAFTING.

A straight line is the shortest distance between two points.

A horizontal line is a line level with the horizon.

A perpendicular line is vertical, or square with a horizontal line.

A base line is a line upon which anything is supposed to stand.

Parallel lines are lines on the same plane, which can never meet, how far soever they may extend.

An angle is the opening between two lines that meet at a corner point.

An angle is denoted by a letter of the alphabet, as A. When three letters are given, as A B C, the middle letter, as B, denotes the angle.

Angles are of three kinds, namely: acute, right, and obtuse.

A right angle forms a square, and the opening contains ninety degrees, and one line is perpendicular to the other.

An obtuse angle is greater than a right angle, and an acute angle is less.

In a right-angled triangle the longest side is called the hypothenuse; it also contains one square corner, or an angle of ninety degrees.

A diagonal is a line extending from corner to corner, dividing a square, or other four-sided figure, into two triangles.

A triangle has three sides and three angles.

The circumference of a circle is a curved line, all points of which are equally distant from the center.

An arc is a portion of a circumference.

A diameter is a straight line meeting the circumference in two points, and passing through the center of a circle.

A chord is a straight line that subtends an arc of a circle.

A radius is half a diameter.

A tangent to a circle is a straight line that has but one point of contact with the circumference of the circle, and forms a right angle with a diameter of the same. To bisect an angle, signifies to divide into two equal parts, by a line passing through it.

A segment of a circle is a portion of the surface between an arc and a chord.

An ellipse, or oval, is an oblong figure bounded by a regular curve.

An equilateral triangle contains three equal sides and three equal angles.

A rectangle is a figure with square corners, but whose sides may be unequal.

The following are some of the polygonal figures that contain equal sides and equal angles:

Name.	Number of sides and angles
Triangle,	3
Square,	4
Pentagon,	5
Hexagon,	6
Heptagon,	7
Octagon,	8
Nonagon,	9
Decagon,	10
Undecagon,	11
Dodecagon.	12

Plate I.

Plate II.

Plate III.

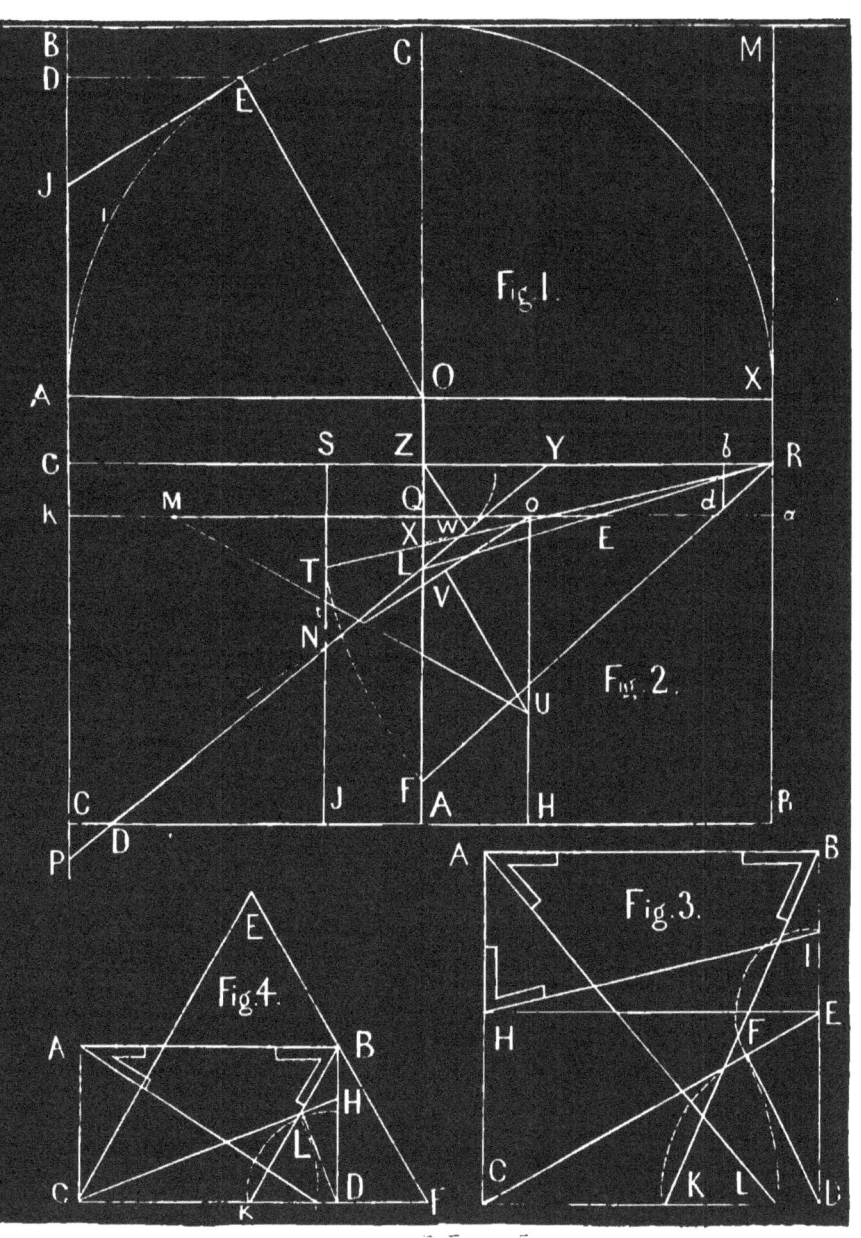

Plate IV.

www.ingramcontent.com/pod-product-compliance
Lightning Source LLC
Chambersburg PA
CBHW020338090426
42735CB00009B/1581